Agricultural Chemistry

Agricultural Chemistry

Edited by
Bert Hudson

Larsen & Keller
www.larsen-keller.com

Agricultural Chemistry
Edited by Bert Hudson
ISBN: 978-1-63549-016-9 (Hardback)

© 2017 Larsen & Keller

☰ Larsen & Keller

Published by Larsen and Keller Education,
5 Penn Plaza,
19th Floor,
New York, NY 10001, USA

Cataloging-in-Publication Data

Agricultural chemistry / edited by Bert Hudson.
    p. cm.
Includes bibliographical references and index.
ISBN 978-1-63549-016-9
1. Agricultural chemistry. 2. Agricultural chemicals.
I. Hudson, Bert.
S585 .A37 2017
668.6--dc23

The publisher's policy is to use permanent paper from mills that operate a sustainable forestry policy. Furthermore, the publisher ensures that the text paper and cover boards used have met acceptable environmental accreditation standards.

Printed and bound in the United States of America.

For more information regarding Larsen and Keller Education and its products, please visit the publisher's website www.larsen-keller.com

Table of Contents

Permissions

Index

Preface

Agricultural chemistry refers to the use of chemistry and biochemistry as they are applied in the field of food production, environmental management and monitoring. It includes the study of the interactions and relationship between plants, bacteria, animals and the surrounding environment. This book explores all the important aspects of agricultural chemistry in the present day scenario. It includes different approaches, evaluations and methodologies on the subject. The topics covered in it offer the readers new insights about this field. It aims to serve as a resource guide for students and facilitates the study of the discipline.

A detailed account of the significant topics covered in this book is provided below:

Chapter 1- The study of chemistry and biochemistry for the use of agricultural production is agricultural chemistry. It is very important for agricultural production and the processing of food products. This chapter will provide an integrated understanding of agricultural chemistry.

Chapter 2- Modern agriculture uses a number of chemicals. Some of these chemicals are pesticide, insecticides, hydroponics, fungicides, bentazon and agricultural lime. Pesticides are used to attract pests and then to destroy them whereas insecticides are used against eggs and larvae to get them killed. This section helps the reader in understanding the reasons for using chemicals in agriculture.

Chapter 3- Fertilizers are applied on plants in order to supply them with more nutrients. They help with the growth of plants and also improve the efficiency of the soil. The chemical compounds that are used in fertilizers are ammonium nitrate, calcium nitrate, potassium nitrate and monocalcium phosphate. This chapter is an overview of the subject matter incorporating all the major aspects of fertilizers.

Chapter 4- The study of the chemical used in the soil is known as soil chemistry. It is affected by many factors; some of these factors are mineral composition, organic matter and the environmental factors. Soil test, soil pH and soil health are the important factors of soil chemistry. This chapter has been written to provide an easy understanding of the varied facets of soil chemistry.

Chapter 5- The nutrition that every plant necessarily needs can be supplied externally also. Nutrient management connects soil, crop and weather to irrigation and soil and water. It is the management of matching the right soil with the right climate and crop management. The section serves as a source to understand the major aspects of plant nutrition.

Chapter 6- Plant hormones help in the growth of the plant, it helps in the regulation of cellular processes in specific cells. Hormones majorly regulate the formation of flowers, stems and leaves. This chapter helps the readers in understanding all the aspects of plant hormones, such as auxin, ethylene, florigen, salicylic acid etc.

It gives me an immense pleasure to thank our entire team for their efforts. Finally in the end, I would like to thank my family and colleagues who have been a great source of inspiration and support.

Editor

Introduction to Agricultural Chemistry

The study of chemistry and biochemistry for the use of agricultural production is agricultural chemistry. It is very important for agricultural production and the processing of foodproducts. This chapter will provide an integrated understanding of agricultural chemistry.

Agricultural Chemistry

Agricultural chemistry is the study of both chemistry and biochemistry which are important in agricultural production, the processing of raw products into foods and beverages, and in environmental monitoring and remediation. These studies emphasize the relationships between plants, animals and bacteria and their environment. The science of chemical compositions and changes involved in the production, protection, and use of crops and livestock. As a basic science, it embraces, in addition to test-tube chemistry, all the life processes through which humans obtain food and fiber for themselves and feed for their animals. As an applied science or technology, it is directed toward control of those processes to increase yields, improve quality, and reduce costs. One important branch of it, chemurgy, is concerned chiefly with utilization of agricultural products as chemical raw materials.

Sciences

The goals of agricultural chemistry are to expand understanding of the causes and effects of biochemical reactions related to plant and animal growth, to reveal opportunities for controlling those reactions, and to develop chemical products that will provide the desired assistance or control. Every scientific discipline that contributes to agricultural progress depends in some way on chemistry. Hence agricultural chemistry is not a distinct discipline, but a common thread that ties together genetics, physiology, microbiology, entomology, and numerous other sciences that impinge on agriculture.

Chemical materials developed to assist in the production of food, feed, and fiber include scores of herbicides, insecticides, fungicides, and other pesticides, plant growth regulators, fertilizers, and animal feed supplements. Chief among these groups from the commercial point of view are manufactured fertilizers, synthetic pesticides (including herbicides), and supplements for feeds. The latter include both nutritional supplements (for example, mineral nutrients) and medicinal compounds for the prevention or control of disease.

Agricultural chemistry often aims at preserving or increasing the fertility of soil, maintaining or improving the agricultural yield, and improving the quality of the crop.

When agriculture is considered with ecology, the sustainablility of an operation is considered. Modern agrochemical industry has gained a reputation for maximising profits while violating sustainable and ecologically viable agricultural principles. Eutrophication, the prevalence of genetically modified crops and the increasing concentration of chemicals in the food chain (e.g. persistent organic pollutants) are only a few consequences of naive industrial agriculture.

History

- In 1761 Johan Gottschalk Wallerius publishes his pioneering work, *Agriculturae fundamenta chemica (Åkerbrukets chemiska grunder)*.

- In 1815 Humphry Davy publishes *Elements of agricultural chemistry*

- In 1842 Justus von Liebig publishes *Animal Chemistry or Organic Chemistry in its applications to Physiology and Pathology.*

- Jöns Jacob Berzelius publishes *Traité de chimie minérale, végétale et animal* (6 vols., 1845–50)

- Jean-Baptiste Boussingault publishes *Agronomie, chimie agricole, et physiologie* (5 vols., 1860–1874; 2nd ed., 1884).

- In 1868 Samuel William Johnson publishes *How Crops Grow*.

- In 1870 S. W. Johnson publishes *How Crops Feed: A treatise on the atmosphere and soil as related to the nutrition of agricultural plants.*

- In 1872 Karl Heinrich Ritthausen publishes *Protein bodies in grains, legumes, and linseed. Contributions to the physiology of seeds for cultivation, nutrition, and fodder*

Chemurgy

Chemurgy is a branch of applied chemistry that is concerned with preparing industrial products from agricultural raw materials. The word "chemurgy" was coined by chemist William J. Hale and first publicized in his 1934 book *The Farm Chemurgic*, the concept was mildly well-developed by the early years of the 20th century. For example, a number of products, including brushes and motion picture film, were made from cellulose. Beginning in the 1920s, some prominent Americans began to advocate a more widespread link between farmers and industry. Among them were William J. Hale and agricultural journalist Wheeler McMillen.

The Hemp Body or Soybean Car

Automaker Henry Ford began to test farm crops for their industrial potential around 1930, and soon settled on hemp and the soybean as particularly promising (the famous Hemp Body or *Soybean Car*). The Ford Motor Company used soybeans in such parts as gearshift knobs and horn buttons, and hemp for the body of the car. The automobile was designed to run on hemp diesel. Ford Motor Company accessed these innovations via the discovery and ingenuity of George Washington Carver, Tuskegee Scientist and Father of Chemurgy.

In 1935, the Farm Chemurgic Council (later renamed the National Farm Chemurgic Council) was formed to encourage greater use of renewable raw materials in industry. In its early years, the Council received substantial publicity. It was perceived by the Roosevelt Administration as a political threat, since Council leaders questioned U.S. Department of Agriculture policies. First placing much of its emphasis on demonstrating the benefits of Agrol (a line of blended motor fuels that included ethanol), the Council drew strong opposition from the petroleum industry. The Agrol pilot plant, which also experienced management and financial difficulties, shut down in 1938. Wheeler McMillen, who had become president of the Council the previous year, decided to distance the chemurgy movement from ethanol, mend fences with the petroleum industry, and place the Council on a more cautious course.

The Council's cause received an unexpected boost when Theodore G. Bilbo, a U.S. senator from Mississippi, sought a means to promote new uses for his region's surplus cotton. To make his goal more politically attractive, he supported a broader research program. In the end, four regional U.S. Department of Agriculture laboratories, dedicated to finding new uses for farm crops, were authorized under the Agricultural Adjustment Act of 1938. The labs were established in Wyndmoor, Pennsylvania; New Orleans, Louisiana; Peoria, Illinois; and Albany, California. Over time, their research agendas expanded, and they became less focused on chemurgy. Nevertheless, their involvement in that field was symbolic of the chemurgy movement's transformation from a cause associated with Roosevelt Administration critics to one with clear support from that administration.

Emergence

Chemurgy demonstrated its worth during World War II, particularly in alleviating the rubber shortage caused when Japan cut off most of America's supply. Corn was used as raw material in much of the synthetic rubber produced during the war. Various other plants, including guayule and kok-saghyz (Russian dandelion), were investigated as rubber sources. In the American Midwest, school children were encouraged to gather milkweed floss, previously considered a nuisance but now valued for a new role as a filler in military life jackets. A priest in Iowa even made news by urging congregants to grow hemp, whose previous reputation as a drug hazard yielded to military requirements for rope and cordage.

Decline

Prospects for chemurgy appeared promising into the 1950s. An article in the December 3, 1951 issue of *Newsweek*, for example, said "the flood of chemurgy seems to be swelling."" But as uses of agricultural raw materials advanced, so did uses for petrochemicals, and non-renewable materials eventually won out in a number of markets. For example, petrochemical detergents were widely used in place of agriculturally derived soaps, and petrochemical plastic wrapping material largely replaced cellophane. The Chemurgic Council went through a period of decline and finally closed its doors in 1977.

In recent years, there has been a resurgence of interest in chemurgy, although the word itself has largely fallen out of usage. In 1990, Wheeler McMillen then 97 years old, addressed a national conference of latter-day chemurgic enthusiasts in Washington, DC. The conference served to launch the New Uses Council, which seeks to further the cause formerly promoted by the Chemurgic Council.

George Washington Carver was one of the most famous scientists of this field. In the Environmental Biography of George Washington Carver titled "My Work is that of Conservation" author Mark D. Hersey writes, "Thus, although he accepted the honorary mantle of "the first and greatest chemurgist," he was hardly in its mainstream. On the contrary, Carver often misconstrued the movement's aims, imagining they fell more in line with his own than in fact they did. Because Carver had devoted his energies to improving the lives of impoverished black farmers, he saw chemurgy as a field in which science addressed "a great human problem." His 1936 injunction to "chemicalize the farm" sprang from his abhorrence of waste rather than a desire for profit, let alone an affinity for chemical pesticides and fertilizers. He wanted "waste products of the farm" to be used for making "insulating boards, paints, dyes, industrial alcohol, plastics of various kinds, rugs, mats and cloth from fiver plants, oils, gums and waxes, etc."

Substitution Examples

- Kenaf for jute (rope)
- castor oil for petroleum-based oil (lubrication)

Agronomy

Agronomy is the science and technology of producing and using plants for food, fuel, fiber, and land reclamation. Agronomy has come to encompass work in the areas of plant genetics, plant physiology, meteorology, and soil science. It is the application of a combination of sciences like biology, chemistry, economics, ecology, earth science, and genetics. Agronomists of today are involved with many issues, including producing food, creating healthier food, managing the environmental impact of agriculture, and extracting energy from plants. Agronomists often specialise in areas such as crop rotation, irrigation and drainage, plant breeding, plant physiology, soil classification, soil fertility, weed control, and insect and pest control.

Plant Breeding

An agronomist field sampling a trial plot of flax.

This area of agronomy involves selective breeding of plants to produce the best crops under various conditions. Plant breeding has increased crop yields and has improved the nutritional value of numerous crops, including corn, soybeans, and wheat. It has also led to the development of new types of plants. For example, a hybrid grain called triticale was produced by crossbreeding rye and wheat. Triticale contains more usable protein than does either rye or wheat. Agronomy has also been instrumental in fruit and vegetable production research.

Biotechnology

Purdue University agronomy professor George Van Scoyoc explains the difference between forest and prairie soils to soldiers of the Indiana National Guard's Agribusiness Development Team at the Beck Agricultural Center in West Lafayette, Indiana

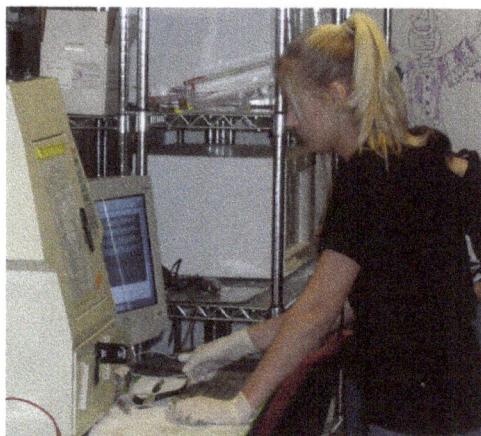

An agronomist mapping a plant genome

Agronomists use biotechnology to extend and expedite the development of desired characteristic. Biotechnology is often a lab activity requiring field testing of the new crop varieties that are developed.

In addition to increasing crop yields agronomic biotechnology is increasingly being applied for novel uses other than food. For example, oilseed is at present used mainly for margarine and other food oils, but it can be modified to produce fatty acids for detergents, substitute fuels and petrochemicals.

Soil Science

Agronomists study sustainable ways to make soils more productive and profitable. They classify soils and analyze them to determine whether they contain nutrients vital to plant growth. Common macronutrients analyzed include compounds of nitrogen, phosphorus, potassium, calcium, magnesium, and sulfur. Soil is also assessed for several micronutrients, like zinc and boron. The percentage of organic matter, soil pH, and nutrient holding capacity (cation exchange capacity) are tested in a regional laboratory. Agronomists will interpret these lab reports and make recommendations to balance soil nutrients for optimal plant growth.

Soil Conservation

In addition, agronomists develop methods to preserve the soil and to decrease the effects of erosion by wind and water. For example, a technique called contour plowing may be used to prevent soil erosion and conserve rainfall. Researchers in agronomy also seek ways to use the soil more effectively in solving other problems. Such problems include the disposal of human and animal manure, water pollution, and pesticide build-up in the soil. Techniques include no-tilling crops, planting of soil-binding grasses along contours on steep slopes, and contour drains of depths up to 1 metre.

Agroecology

Agroecology is the management of agricultural systems with an emphasis on ecological and environmental perspectives. This area is closely associated with work in the areas of sustainable agriculture, organic farming, and alternative food systems and the development of alternative cropping systems.

Theoretical Modeling

Theoretical production ecology tries to quantitatively study the growth of crops. The plant is treated as a kind of biological factory, which processes light, carbon dioxide, water, and nutrients into harvestable parts. Main parameters kept into consideration are temperature, sunlight, standing crop biomass, plant production distribution, nutrient and water supply.

Chemicals and Techniques Related to Agricultural Chemistry

Modern agriculture uses a number of chemicals. Some of these chemicals are pesticide, insecticides, hydroponics, fungicides, bentazon and agricultural lime. Pesticides are used to attract pests and then to destroy them whereas insecticides are used against eggs and larvae to get them killed. This section helps the reader in understanding the reasons for using chemicals in agriculture.

Pesticide

A crop-duster spraying pesticide on a field

Pesticides are substances meant for attracting, seducing, and then destroying any pest. They are a class of biocide. The most common use of pesticides is as plant protection products (also known as crop protection products), which in general protect plants from damaging influences such as weeds, fungi, or insects. This use of pesticides is so common that the term *pesticide* is often treated as synonymous with *plant protection product*, although it is in fact a broader term, as pesticides are also used for non-agricultural purposes. The term pesticide includes all of the following: herbicide, insecticide, insect growth regulator, nematicide, termiticide, molluscicide, piscicide, avicide, rodenticide, predacide, bactericide, insect repellent, animal repellent, antimicrobial, fungicide, disinfectant (antimicrobial), and sanitizer.

A Lite-Trac four-wheeled self-propelled crop sprayer spraying pesticide on a field

In general, a pesticide is a chemical or biological agent (such as a virus, bacterium, antimicrobial, or disinfectant) that deters, incapacitates, kills, or otherwise discourages pests. Target pests can include insects, plant pathogens, weeds, mollusks, birds, mammals, fish, nematodes (roundworms), and microbes that destroy property, cause nuisance, or spread disease, or are disease vectors. Although pesticides have benefits, some also have drawbacks, such as potential toxicity to humans and other species. According to the Stockholm Convention on Persistent Organic Pollutants, 9 of the 12 most dangerous and persistent organic chemicals are organochlorine pesticides.

Definition

The Food and Agriculture Organization (FAO) has defined *pesticide* as:

> any substance or mixture of substances intended for preventing, destroying, or controlling any pest, including vectors of human or animal disease, unwanted species of plants or animals, causing harm during or otherwise interfering with the production, processing, storage, transport, or marketing of food, agricultural commodities, wood and wood products or animal feedstuffs, or substances that may be administered to animals for the control of insects, arachnids, or other pests in or on their bodies. The term includes substances intended for use as a plant growth regulator, defoliant, desiccant, or agent for thinning fruit or preventing the premature fall of fruit. Also used as substances applied to crops either before or after harvest to protect the commodity from deterioration during storage and transport.

Pesticides can be classified by target organism, chemical structure (e.g., organic, inorganic, synthetic, or biological (biopesticide), although the distinction can sometimes blur), and physical state (e.g. gaseous (fumigant)). Biopesticides include microbial pesticides and biochemical pesticides. Plant-derived pesticides, or "botanicals", have been developing quickly. These include the pyrethroids, rotenoids, nicotinoids, and a fourth group that includes strychnine and scilliroside.

Many pesticides can be grouped into chemical families. Prominent insecticide families include organochlorines, organophosphates, and carbamates. Organochlorine hydrocarbons (e.g., DDT) could be separated into dichlorodiphenylethanes, cyclodiene compounds, and other related compounds. They operate by disrupting the sodium/potassium balance of the nerve fiber, forcing the nerve to transmit continuously. Their toxicities vary greatly, but they have been phased out because of their persistence and potential to bioaccumulate. Organophosphate and carbamates largely replaced organochlorines. Both operate through inhibiting the enzyme acetylcholinesterase, allowing acetylcholine to transfer nerve impulses indefinitely and causing a variety of symptoms such as weakness or paralysis. Organophosphates are quite toxic to vertebrates, and have in some cases been replaced by less toxic carbamates. Thiocarbamate and dithiocarbamates are subclasses of carbamates. Prominent families of herbicides include phenoxy and benzoic acid herbicides (e.g. 2,4-D), triazines (e.g., atrazine), ureas (e.g., diuron), and Chloroacetanilides (e.g., alachlor). Phenoxy compounds tend to selectively kill broad-leaf weeds rather than grasses. The phenoxy and benzoic acid herbicides function similar to plant growth hormones, and grow cells without normal cell division, crushing the plant's nutrient transport system. Triazines interfere with photosynthesis. Many commonly used pesticides are not included in these families, including glyphosate.

Pesticides can be classified based upon their biological mechanism function or application method. Most pesticides work by poisoning pests. A systemic pesticide moves inside a plant following absorption by the plant. With insecticides and most fungicides, this movement is usually upward (through the xylem) and outward. Increased efficiency may be a result. Systemic insecticides, which poison pollen and nectar in the flowers, may kill bees and other needed pollinators.

In 2009, the development of a new class of fungicides called paldoxins was announced. These work by taking advantage of natural defense chemicals released by plants called phytoalexins, which fungi then detoxify using enzymes. The paldoxins inhibit the fungi's detoxification enzymes. They are believed to be safer and greener.

Uses

Pesticides are used to control organisms that are considered to be harmful. For example, they are used to kill mosquitoes that can transmit potentially deadly diseases like West Nile virus, yellow fever, and malaria. They can also kill bees, wasps or ants that can cause allergic reactions. Insecticides can protect animals from illnesses that can be caused by parasites such as fleas. Pesticides can prevent sickness in humans that could be caused by moldy food or diseased produce. Herbicides can be used to clear roadside weeds, trees and brush. They can also kill invasive weeds that may cause environmental damage. Herbicides are commonly applied in ponds and lakes to control algae and plants such as water grasses that can interfere with activities like swimming and fishing and cause the water to look or smell unpleasant. Uncontrolled pests such as termites and mold can damage structures such as houses. Pesticides are used in grocery stores and food storage facilities to manage rodents and insects that infest food such as grain. Each use of a pesticide carries some associated risk. Proper pesticide use decreases these associated risks to a level deemed acceptable by pesticide regulatory agencies such as the United States Environmental Protection Agency (EPA) and the Pest Management Regulatory Agency (PMRA) of Canada.

DDT, sprayed on the walls of houses, is an organochlorine that has been used to fight malaria since the 1950s. Recent policy statements by the World Health Organization have given stronger support to this approach. However, DDT and other organochlorine pesticides have been banned in most countries worldwide because of their persistence in the environment and human toxicity. DDT use is not always effective, as resistance to DDT was identified in Africa as early as 1955, and by 1972 nineteen species of mosquito worldwide were resistant to DDT.

Amount used

In 2006 and 2007, the world used approximately 2.4 megatonnes (5.3×10^9 lb) of pesticides, with herbicides constituting the biggest part of the world pesticide use at 40%, followed by insecticides (17%) and fungicides (10%). In 2006 and 2007 the U.S. used approximately 0.5 megatonnes (1.1×10^9 lb) of pesticides, accounting for 22% of the world total, including 857 million pounds (389 kt) of conventional pesticides, which are used in the agricultural sector (80% of conventional pesticide use) as well as the industrial, commercial, governmental and home & garden sectors. Pesticides are also found in majority of U.S. households with 78 million out of the 105.5 million households indicating that they use some form of pesticide. As of 2007, there were more than 1,055 active ingredients registered as pesticides, which yield over 20,000 pesticide products that are marketed in the United States.

The US used some 1 kg (2.2 pounds) per hectare of arable land compared with: 4.7 kg in China, 1.3 kg in the UK, 0.1 kg in Cameroon, 5.9 kg in Japan and 2.5 kg in Italy. Insecticide use in the US has declined by more than half since 1980, (.6%/yr) mostly due to the near phase-out of organophosphates. In corn fields, the decline was even steeper, due to the switchover to transgenic Bt corn.

For the global market of crop protection products, market analysts forecast revenues of over 52 billion US$ in 2019.

Benefits

Pesticides can save farmers' money by preventing crop losses to insects and other pests; in the U.S., farmers get an estimated fourfold return on money they spend on pesticides. One study found that not using pesticides reduced crop yields by about 10%. Another study, conducted in 1999, found that a ban on pesticides in the United States may result in a rise of food prices, loss of jobs, and an increase in world hunger.

There are two levels of benefits for pesticide use, primary and secondary. Primary benefits are direct gains from the use of pesticides and secondary benefits are effects that are more long-term.

Primary Benefits

1. Controlling pests and plant disease vectors

    o   Improved crop/livestock yields

    o   Improved crop/livestock quality

    o   Invasive species controlled

2. Controlling human/livestock disease vectors and nuisance organisms

    o   Human lives saved and suffering reduced

    o   Animal lives saved and suffering reduced

    o   Diseases contained geographically

3. Controlling organisms that harm other human activities and structures

    o   Drivers view unobstructed

    o   Tree/brush/leaf hazards prevented

    o   Wooden structures protected

Monetary

Every dollar ($1) that is spent on pesticides for crops yields four dollars ($4) in crops saved. This means based that, on the amount of money spent per year on pesticides, $10 billion, there is an additional $40 billion savings in crop that would be lost due to damage by insects and weeds. In general, farmers benefit from having an increase in crop yield and from being able to grow a variety of crops throughout the year. Consumers of agricultural products also benefit from being able to afford the vast quantities of produce available year-round. The general public also benefits from the use of pesticides for the control of insect-borne diseases and illnesses, such as malaria. The use of pesticides creates a large job market within the agri-chemical sector.

Costs

On the cost side of pesticide use there can be costs to the environment, costs to human health, as well as costs of the development and research of new pesticides.

Health Effects

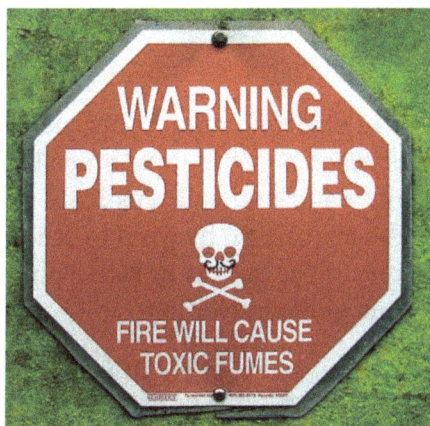

A sign warning about potential pesticide exposure.

Pesticides may cause acute and delayed health effects in people who are exposed. Pesticide exposure can cause a variety of adverse health effects, ranging from simple irritation of the skin and eyes to more severe effects such as affecting the nervous system, mimicking hormones causing reproductive problems, and also causing cancer. A 2007 systematic review found that "most studies on non-Hodgkin lymphoma and leukemia showed positive associations with pesticide exposure" and thus concluded that cosmetic use of pesticides should be decreased. There is substantial evidence of associations between organophosphate insecticide exposures and neurobehavioral alterations. Limited evidence also exists for other negative outcomes from pesticide exposure including neurological, birth defects, and fetal death.

The American Academy of Pediatrics recommends limiting exposure of children to pesticides and using safer alternatives:

The World Health Organization and the UN Environment Programme estimate that each year, 3 million workers in agriculture in the developing world experience severe poisoning from pesticides, about 18,000 of whom die. Owing to inadequate regulation and safety precautions, 99% of pesticide related deaths occur in developing countries that account for only 25% of pesticide usage. According to one study, as many as 25 million workers in developing countries may suffer mild pesticide poisoning yearly.

There are several careers aside from agriculture that may also put individuals at risk of health effects from pesticide exposure including pet groomers, groundskeepers, and fumigators.

One study found pesticide self-poisoning the method of choice in one third of suicides worldwide, and recommended, among other things, more restrictions on the types of pesticides that are most harmful to humans.

A 2014 epidemiological review found associations between autism and exposure to certain pesticides, but noted that the available evidence was insufficient to conclude that the relationship was causal.

Environmental Effect

Pesticide use raises a number of environmental concerns. Over 98% of sprayed insecticides and 95% of herbicides reach a destination other than their target species, including non-target species, air, water and soil. Pesticide drift occurs when pesticides suspended in the air as particles are carried by wind to other areas, potentially contaminating them. Pesticides are one of the causes of water pollution, and some pesticides are persistent organic pollutants and contribute to soil contamination.

In addition, pesticide use reduces biodiversity, contributes to pollinator decline, destroys habitat (especially for birds), and threatens endangered species. Pests can develop a resistance to the pesticide (pesticide resistance), necessitating a new pesticide. Alternatively a greater dose of the pesticide can be used to counteract the resistance, although this will cause a worsening of the ambient pollution problem.

Since chlorinated hydrocarbon pesticides dissolve in fats and are not excreted, organisms tend to retain them almost indefinitely. Biological magnification is the process whereby these chlorinated hydrocarbons (pesticides) are more concentrated at each level of the food chain. Among marine animals, pesticide concentrations are higher in carnivorous fishes, and even more so in the fish-eating birds and mammals at the top of the ecological pyramid. Global distillation is the process whereby pesticides are transported from warmer to colder regions of the Earth, in particular the Poles and mountain tops. Pesticides that evaporate into the atmosphere at relatively high temperature can be carried considerable distances (thousands of kilometers) by the wind to an area of lower temperature, where they condense and are carried back to the ground in rain or snow.

In order to reduce negative impacts, it is desirable that pesticides be degradable or at least quickly deactivated in the environment. Such loss of activity or toxicity of pesticides is due to both innate chemical properties of the compounds and environmental processes or conditions. For example, the presence of halogens within a chemical structure often slows down degradation in an aerobic environment. Adsorption to soil may retard pesticide movement, but also may reduce bioavailability to microbial degraders.

Economics

Human health and environmental cost from pesticides in the United States is estimated at $9.6 billion offset by about $40 billion in increased agricultural production:

Additional costs include the registration process and the cost of purchasing pesticides. The registration process can take several years to complete (there are 70 different types of field test) and can cost $50–70 million for a single pesticide. Annually the United States spends $10 billion on pesticides.

Alternatives

Alternatives to pesticides are available and include methods of cultivation, use of biological pest controls (such as pheromones and microbial pesticides), genetic engineering, and methods of interfering with insect breeding. Application of composted yard waste has also been used as a way of controlling pests. These methods are becoming increasingly popular and often are safer than traditional chemical pesticides. In addition, EPA is registering reduced-risk conventional pesticides in increasing numbers.

Cultivation practices include polyculture (growing multiple types of plants), crop rotation, planting crops in areas where the pests that damage them do not live, timing planting according to when pests will be least problematic, and use of trap crops that attract pests away from the real crop. In the U.S., farmers have had success controlling insects by spraying with hot water at a cost that is about the same as pesticide spraying.

Release of other organisms that fight the pest is another example of an alternative to pesticide use. These organisms can include natural predators or parasites of the pests. Biological pesticides based on entomopathogenic fungi, bacteria and viruses cause disease in the pest species can also be used.

Interfering with insects' reproduction can be accomplished by sterilizing males of the target species and releasing them, so that they mate with females but do not produce offspring. This technique was first used on the screwworm fly in 1958 and has since been used with the medfly, the tsetse fly, and the gypsy moth. However, this can be a costly, time consuming approach that only works on some types of insects.

Agroecology emphasize nutrient recycling, use of locally available and renewable resources, adaptation to local conditions, utilization of microenvironments, reliance on indigenous knowledge and yield maximization while maintaining soil productivity. Agroecology also emphasizes empowering people and local communities to contribute to development, and encouraging "multi-directional" communications rather than the conventional "top-down" method.

Push Pull Strategy

The term "push-pull" was established in 1987 as an approach for integrated pest management (IPM). This strategy uses a mixture of behavior-modifying stimuli to manipulate the distribution and abundance of insects. "Push" means the insects are repelled or deterred away from whatever resource that is being protected. "Pull" means that certain stimuli (semiochemical stimuli, pheromones, food additives, visual stimuli, genetically altered plants, etc.) are used to attract pests to trap crops where they will be killed. There are numerous different components involved in order to implement a Push-Pull Strategy in IPM.

Many case studies testing the effectiveness of the push-pull approach have been done across the world. The most successful push-pull strategy was developed in Africa for subsistence farming. Another successful case study was performed on the control of *Helicoverpa* in cotton crops in Australia. In Europe, the Middle East, and the United States, push-pull strategies were successfully used in the controlling of *Sitona lineatus* in bean fields.

Some advantages of using the push-pull method are less use of chemical or biological materials and better protection against insect habituation to this control method. Some disadvantages of the push-pull strategy is that if there is a lack of appropriate knowledge of behavioral and chemical ecology of the host-pest interactions then this method becomes unreliable. Furthermore, because the push-pull method is not a very popular method of IPM operational and registration costs are higher.

Effectiveness

Some evidence shows that alternatives to pesticides can be equally effective as the use of chemicals. For example, Sweden has halved its use of pesticides with hardly any reduction in crops. In Indonesia, farmers have reduced pesticide use on rice fields by 65% and experienced a 15% crop increase. A study of Maize fields in northern Florida found that the application of composted yard waste with high carbon to nitrogen ratio to agricultural fields was highly effective at reducing the population of plant-parasitic nematodes and increasing crop yield, with yield increases ranging from 10% to 212%; the observed effects were long-term, often not appearing until the third season of the study.

However, pesticide resistance is increasing. In the 1940s, U.S. farmers lost only 7% of their crops to pests. Since the 1980s, loss has increased to 13%, even though more pesticides are being used. Between 500 and 1,000 insect and weed species have developed pesticide resistance since 1945.

Types

Pesticides are often referred to according to the type of pest they control. Pesticides can also be considered as either biodegradable pesticides, which will be broken down by microbes and other living beings into harmless compounds, or persistent pesticides, which may take months or years before they are broken down: it was the persistence of DDT, for example, which led to its accumulation in the food chain and its killing of birds of prey at the top of the food chain. Another way to think about pesticides is to consider those that are chemical pesticides or are derived from a common source or production method.

Some examples of chemically-related pesticides are:

Organophosphate Pesticides

Organophosphates affect the nervous system by disrupting, acetylcholinesterase activity, the enzyme that regulates acetylcholine, a neurotransmitter. Most organophosphates are insecticides. They were developed during the early 19th century, but their effects on insects, which are similar to their effects on humans, were discovered in 1932. Some are very poisonous. However, they usually are not persistent in the environment.

Carbamate Pesticides

Carbamate pesticides affect the nervous system by disrupting an enzyme that regulates acetylcholine, a neurotransmitter. The enzyme effects are usually reversible. There are several subgroups within the carbamates.

Organochlorine Insecticides

They were commonly used in the past, but many have been removed from the market due to their health and environmental effects and their persistence (e.g., DDT, chlordane, and toxaphene).

Pyrethroid Pesticides

They were developed as a synthetic version of the naturally occurring pesticide pyrethrin, which is found in chrysanthemums. They have been modified to increase their stability in the environment. Some synthetic pyrethroids are toxic to the nervous system.

Sulfonylurea Herbicides

The following sulfonylureas have been commercialized for weed control: amidosulfuron, azimsulfuron, bensulfuron-methyl, chlorimuron-ethyl, ethoxysulfuron, flazasulfuron, flupyrsulfuron-methyl-sodium, halosulfuron-methyl, imazosulfuron,nicosulfuron, oxasulfuron, primisulfuron-methyl, pyrazosulfuron-ethyl, rimsulfuron,sulfometuron-methyl Sulfosulfuron, terbacil, bispyribac-sodium, cyclosulfamuron, and pyrithiobac-sodium. Nicosulfuron, triflusulfuron methyl, andchlorsulfuron are broad-spectrum herbicides that kill plants by inhibiting the enzyme acetolactate synthase. In the 1960s, more than 1 kg/ha (0.89 lb/acre) crop protection chemical was typically applied, while sulfonylureates allow as little as 1% as much material to achieve the same effect.

Biopesticides

Biopesticides are certain types of pesticides derived from such natural materials as animals, plants, bacteria, and certain minerals. For example, canola oil and baking soda have pesticidal applications and are considered biopesticides. Biopesticides fall into three major classes:

- Microbial pesticides which consist of bacteria, entomopathogenic fungi or viruses (and sometimes includes the metabolites that bacteria or fungi produce). Entomopathogenic nematodes are also often classed as microbial pesticides, even though they are multi-cellular.

- Biochemical pesticides or herbal pesticides are naturally occurring substances that control (or monitor in the case of pheromones) pests and microbial diseases.

- Plant-incorporated protectants (PIPs) have genetic material from other species incorporated into their genetic material (*i.e.* GM crops). Their use is controversial, especially in many European countries.

Classified By Type of Pest

Pesticides that are related to the type of pests are:

Further Types of Pesticides

The term pesticide also include these substances:

Defoliants : Cause leaves or other foliage to drop from a plant, usually to facilitate harvest.

Desiccants : Promote drying of living tissues, such as unwanted plant tops.

Insect growth regulators : Disrupt the molting, maturity from pupal stage to adult, or other life processes of insects.

Plant growth regulators : Substances (excluding fertilizers or other plant nutrients) that alter the expected growth, flowering, or reproduction rate of plants.

Regulation

International

In most countries, pesticides must be approved for sale and use by a government agency.

In Europe, recent EU legislation has been approved banning the use of highly toxic pesticides including those that are carcinogenic, mutagenic or toxic to reproduction, those that are endocrine-disrupting, and those that are persistent, bioaccumulative and toxic

(PBT) or very persistent and very bioaccumulative (vPvB). Measures were approved to improve the general safety of pesticides across all EU member states.

Though pesticide regulations differ from country to country, pesticides, and products on which they were used are traded across international borders. To deal with inconsistencies in regulations among countries, delegates to a conference of the United Nations Food and Agriculture Organization adopted an International Code of Conduct on the Distribution and Use of Pesticides in 1985 to create voluntary standards of pesticide regulation for different countries. The Code was updated in 1998 and 2002. The FAO claims that the code has raised awareness about pesticide hazards and decreased the number of countries without restrictions on pesticide use.

Three other efforts to improve regulation of international pesticide trade are the United Nations London Guidelines for the Exchange of Information on Chemicals in International Trade and the United Nations Codex Alimentarius Commission. The former seeks to implement procedures for ensuring that prior informed consent exists between countries buying and selling pesticides, while the latter seeks to create uniform standards for maximum levels of pesticide residues among participating countries. Both initiatives operate on a voluntary basis.

Pesticides safety education and pesticide applicator regulation are designed to protect the public from pesticide misuse, but do not eliminate all misuse. Reducing the use of pesticides and choosing less toxic pesticides may reduce risks placed on society and the environment from pesticide use. Integrated pest management, the use of multiple approaches to control pests, is becoming widespread and has been used with success in countries such as Indonesia, China, Bangladesh, the U.S., Australia, and Mexico. IPM attempts to recognize the more widespread impacts of an action on an ecosystem, so that natural balances are not upset. New pesticides are being developed, including biological and botanical derivatives and alternatives that are thought to reduce health and environmental risks. In addition, applicators are being encouraged to consider alternative controls and adopt methods that reduce the use of chemical pesticides.

Pesticides can be created that are targeted to a specific pest's lifecycle, which can be environmentally more friendly. For example, potato cyst nematodes emerge from their protective cysts in response to a chemical excreted by potatoes; they feed on the potatoes and damage the crop. A similar chemical can be applied to fields early, before the potatoes are planted, causing the nematodes to emerge early and starve in the absence of potatoes.

United States

In the United States, the Environmental Protection Agency (EPA) is responsible for regulating pesticides under the Federal Insecticide, Fungicide, and Rodenticide Act (FIFRA) and the Food Quality Protection Act (FQPA). Studies must be conducted to establish the conditions in which the material is safe to use and the effectiveness against the intended pest(s). The EPA regulates pesticides to ensure that these products do not pose adverse effects to humans or the environment. Pesticides produced before November 1984 continue to be reassessed in order to meet the current scientific and regulatory standards. All registered pesticides are reviewed every 15 years to ensure they meet the proper standards. During the registration process, a label is created. The label contains directions for proper use of the material in addition to safety restrictions. Based on acute toxicity, pesticides are assigned to a Toxicity Class.

Preparation for an application of hazardous herbicide in USA.

Some pesticides are considered too hazardous for sale to the general public and are designated restricted use pesticides. Only certified applicators, who have passed an exam, may purchase or supervise the application of restricted use pesticides. Records of sales and use are required to be maintained and may be audited by government agencies charged with the enforcement of pesticide regulations. These records must be made available to employees and state or territorial environmental regulatory agencies.

The EPA regulates pesticides under two main acts, both of which amended by the Food Quality Protection Act of 1996. In addition to the EPA, the United States Department of Agriculture (USDA) and the United States Food and Drug Administration (FDA) set standards for the level of pesticide residue that is allowed on or in crops. The EPA looks at what the potential human health and environmental effects might be associated with the use of the pesticide.

In addition, the U.S. EPA uses the National Research Council's four-step process for human health risk assessment: (1) Hazard Identification, (2) Dose-Response Assessment, (3) Exposure Assessment, and (4) Risk Characterization.

Recently Kaua'i County (Hawai'i) passed Bill No. 2491 to add an article to Chapter 22 of the county's code relating to pesticides and GMOs. The bill strengthens protections of local communities in Kaua'i where many large pesticide companies test their products.

History

Since before 2000 BC, humans have utilized pesticides to protect their crops. The first known pesticide was elemental sulfur dusting used in ancient Sumer about 4,500 years ago in ancient Mesopotamia. The Rig Veda, which is about 4,000 years old, mentions the use of poisonous plants for pest control. By the 15th century, toxic chemicals such as arsenic, mercury, and lead were being applied to crops to kill pests. In the 17th century, nicotine sulfate was extracted from tobacco leaves for use as an insecticide. The 19th century saw the introduction of two more natural pesticides, pyrethrum, which is derived from chrysanthemums, and rotenone, which is derived from the roots of tropical vegetables. Until the 1950s, arsenic-based pesticides were dominant. Paul Müller discovered that DDT was a very effective insecticide. Organochlorines such as DDT were dominant, but they were replaced in the U.S. by organophosphates and carbamates by 1975. Since then, pyrethrin compounds have become the dominant insecticide. Herbicides became common in the 1960s, led by "triazine and other nitrogen-based compounds, carboxylic acids such as 2,4-dichlorophenoxyacetic acid, and glyphosate".

The first legislation providing federal authority for regulating pesticides was enacted in 1910; however, decades later during the 1940s manufacturers began to produce large amounts of synthetic pesticides and their use became widespread. Some sources consider the 1940s and 1950s to have been the start of the "pesticide era." Although the U.S. Environmental Protection Agency was established in 1970 and amendments to the pesticide law in 1972, pesticide use has increased 50-fold since 1950 and 2.3 million tonnes (2.5 million short tons) of industrial pesticides are now used each year. Seventy-five percent of all pesticides in the world are used in developed countries, but use in developing countries is increasing. A study of USA pesticide use trends through 1997 was published in 2003 by the National Science Foundation's Center for Integrated Pest Management.

In the 1960s, it was discovered that DDT was preventing many fish-eating birds from reproducing, which was a serious threat to biodiversity. Rachel Carson wrote the best-selling book *Silent Spring* about biological magnification. The agricultural use of DDT is now banned under the Stockholm Convention on Persistent Organic Pollutants, but it is still used in some developing nations to prevent malaria and other tropical diseases by spraying on interior walls to kill or repel mosquitoes.

Insecticide

An insecticide is a substance used to kill insects. They include ovicides and larvicides used against insect eggs and larvae, respectively. Insecticides are used in agriculture, medicine, industry and by consumers. Insecticides are claimed to be a major factor behind the increase in agricultural 20th century's productivity. Nearly all insecticides have the potential to significantly alter ecosystems; many are toxic to humans; some concentrate along the food chain.

Insecticides can be classified in two major groups: systemic insecticides, which have residual or long term activity; and contact insecticides, which have no residual activity.

FLIT manual spray pump for insecticides from 1928

Furthermore, one can distinguish three types of insecticide. 1. Natural insecticides, such as nicotine, pyrethrum and neem extracts, made by plants as defenses against insects. 2. Inorganic insecticides, which are metals. 3. Organic insecticides, which are organic chemical compounds, mostly working by contact.

The mode of action describes how the pesticide kills or inactivates a pest. It provides another way of classifying insecticides. Mode of action is important in understanding whether an insecticide will be toxic to unrelated species, such as fish, birds and mammals.

Insecticides are distinct from insect repellents, which do not kill.

Type of Activity

Systemic insecticides become incorporated and distributed systemically throughout the whole plant. When insects feed on the plant, they ingest the insecticide. Systemic insecticides produced by transgenic plants are called plant-incorporated protectants (PIPs). For instance, a gene that codes for a specific Bacillus thuringiensis biocidal protein was introduced into corn and other species. The plant manufactures the protein, which kills the insect when consumed.

Contact insecticides are toxic to insects upon direct contact. These can be inorganic insecticides, which are metals and include arsenates, copper and fluorine compounds, which are less commonly used, and the commonly used sulfur. Contact insecticides can be organic insecticides, i.e. organic chemical compounds, synthetically produced, and comprising the largest numbers of pesticides used today. Or they can be natural compounds like pyrethrum, neem oil etc. Contact insecticides usually have no residual activity.

Efficacy can be related to the quality of pesticide application, with small droplets, such as aerosols often improving performance.

Biological Pesticides

Many organic compounds are produced by plants for the purpose of defending the host plant from predation. A trivial case is tree rosin, which is a natural insecticide. Specific, the production of oleoresin by conifer species is a component of the defense response against insect attack and fungal pathogen infection. Many fragrances, e.g. oil of wintergreen, are in fact antifeedants.

Four extracts of plants are in commercial use: pyrethrum, rotenone, neem oil, and various essential oils

Other Biological Approaches

Plant-incorporated Protectants

Transgenic crops that act as insecticides began in 1996 with a genetically modified potato that produced the Cry protein, derived from the bacterium Bacillus thuringiensis, which is toxic to beetle larvae such as the Colorado potato beetle. The technique has been expanded to include the use of RNA interference RNAi that fatally silences crucial insect genes. RNAi likely evolved as a defense against viruses. Midgut cells in many larvae take up the molecules and help spread the signal. The technology can target only insects that have the silenced sequence, as was demonstrated when a particular RNAi affected only one of four fruit fly species. The technique is expected to replace many other insecticides, which are losing effectiveness due to the spread of pesticide resistance.

Enzymes

Many plants exude substances to repel insects. Premier examples are substances activated by the enzyme myrosinase. This enzyme converts glucosinolates to various compounds that are toxic to herbivorous insects. One product of this enzyme is allyl isothiocyanate, the pungent ingredient in horseradish sauces.

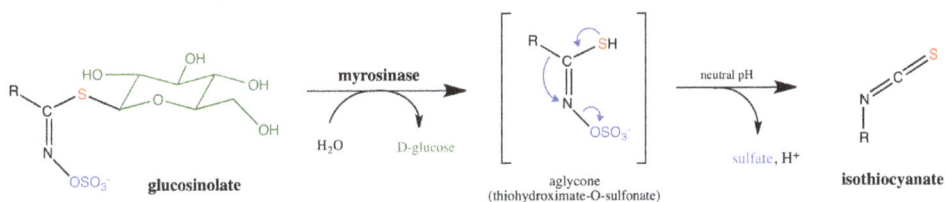

Biosynthesis of antifeedants by the action of myrosinase.

The myrosinase is released only upon crushing the flesh of horseradish. Since allyl isothiocyanate is harmful to the plant as well as the insect, it is stored in the harmless form of the glucosinolate, separate from the myrosinase enzyme.

Bacterial

Bacillus thuringiensis is a bacterial disease that affects Lepidopterans and some other insects. Toxins produced by strains of this bacterium are used as a larvicide against caterpillars, beetles, and mosquitoes. Toxins from *Saccharopolyspora spinosa* are isolated from fermentations and sold as Spinosad. Because these toxins have little effect on other organisms, they are considered more environmentally friendly than synthetic pesticides. The toxin from *B. thuringiensis* (Bt toxin) has been incorporated directly into plants through the use of genetic engineering. Other biological insecticides include products based on entomopathogenic fungi (e.g., *Beauveria bassiana*, *Metarhizium anisopliae*), nematodes (e.g., *Steinernema feltiae*) and viruses (e.g., *Cydia pomonella* granulovirus).

Synthetic Insecticide

A major emphasis of organic chemistry is the development of chemical tools to enhance agricultural productivity. Insecticides represent a major area of emphasis. Many of the major insecticides are inspired by biological analogues. Many others are completely alien to nature.

Organochlorides

The best known organochloride, DDT, was created by Swiss scientist Paul Müller. For this discovery, he was awarded the 1948 Nobel Prize for Physiology or Medicine. DDT was introduced in 1944. It functions by opening sodium channels in the insect's nerve cells. The contemporaneous rise of the chemical industry facilitated large-scale production of DDT and related chlorinated hydrocarbons.

Organophosphates and Carbamates

Organophosphates are another large class of contact insecticides. These also target the insect's nervous system. Organophosphates interfere with the enzymes acetylcholinesterase and other cholinesterases, disrupting nerve impulses and killing or disabling the insect. Organophosphate insecticides and chemical warfare nerve agents (such as sarin, tabun, soman, and VX) work in the same way. Organophosphates have a cumulative toxic effect to wildlife, so multiple exposures to the chemicals amplifies the toxicity. In the US, organophosphate use declined with the rise of substitutes.

Carbamate insecticides have similar mechanisms to organophosphates, but have a much shorter duration of action and are somewhat less toxic.

Pyrethroids

Pyrethroid pesticides mimic the insecticidal activity of the natural compound pyrethrum, the biopesticide found in pyrethrins. These compounds are nonpersistent sodium channel modulators and are less toxic than organophosphates and carbamates. Compounds in this group are often applied against household pests.

Neonicotinoids

Neonicotinoids are synthetic analogues of the natural insecticide nicotine (with much lower acute mammalian toxicity and greater field persistence). These chemicals are acetylcholine receptor agonists. They are broad-spectrum systemic insecticides, with rapid action (minutes-hours). They are applied as sprays, drenches, seed and soil treatments. Treated insects exhibit leg tremors, rapid wing motion, stylet withdrawal (aphids), disoriented movement, paralysis and death. Imidacloprid may be the most common. It has recently come under scrutiny for allegedly pernicious effects on honeybees and its potential to increase the susceptibility of rice to planthopper attacks.

Ryanoids

Ryanoids are synthetic analogues with the same mode of action as ryanodine, a naturally occurring insecticide extracted from *Ryania speciosa* (Flacourtiaceae). They bind to calcium channels in cardiac and skeletal muscle, blocking nerve transmission. Only one such insecticide is currently registered, Rynaxypyr, generic name chlorantraniliprole.

Insect Growth Regulators

Insect growth regulator (IGR) is a term coined to include insect hormone mimics and an earlier class of chemicals, the benzoylphenyl ureas, which inhibit chitin(exoskeleton) biosynthesis in insects. Diflubenzuron is a member of the latter class, used primarily to control caterpillars that are pests. The most successful insecticides in this class are the juvenoids (juvenile hormone analogues). Of these, methoprene is most widely used. It has no observable acute toxicity in rats and is approved by World Health Organization (WHO) for use in drinking water cisterns to combat malaria. Most of its uses are to combat insects where the adult is the pest, including mosquitoes, several fly species, and fleas. Two very similar products, hydroprene and kinoprene, are used for controlling species such as cockroaches and white flies. Methoprene was registered with the EPA in 1975. Virtually no reports of resistance have been filed. A more recent type of IGR is the ecdysone agonist tebufenozide (MIMIC), which is used in forestry and other applications for control of caterpillars, which are far more sensitive to its hormonal effects than other insect orders.

Environmental Effects

Effects on Nontarget Species

Some insecticides kill or harm other creatures in addition to those they are intended to kill. For example, birds may be poisoned when they eat food that was recently sprayed with insecticides or when they mistake an insecticide granule on the ground for food and eat it.

Sprayed insecticide may drift from the area to which it is applied and into wildlife areas, especially when it is sprayed aerially.

DDT

The development of DDT was motivated by desire to replace more dangerous or less effective alternatives. DDT was introduced to replace lead and arsenic-based compounds, which were in widespread use in the early 1940s.

DDT was brought to public attention by Rachel Carson's book *Silent Spring*. One side-effect of DDT is to reduce the thickness of shells on the eggs of predatory birds. The shells sometimes become too thin to be viable, reducing bird populations. This occurs with DDT and related compounds due to the process of bioaccumulation, wherein the chemical, due to its stability and fat solubility, accumulates in organisms' fatty tissues. Also, DDT may biomagnify, which causes progressively higher concentrations in the body fat of animals farther up the food chain. The near-worldwide ban on agricultural use of DDT and related chemicals has allowed some of these birds, such as the peregrine falcon, to recover in recent years. A number of organochlorine pesticides have been banned from most uses worldwide. Globally they are controlled via the Stockholm Convention on persistent organic pollutants. These include: aldrin, chlordane, DDT, dieldrin, endrin, heptachlor, mirex and toxaphene.

Pollinator Decline

Insecticides can kill bees and may be a cause of pollinator decline, the loss of bees that pollinate plants, and colony collapse disorder (CCD), in which worker bees from a beehive or Western honey bee colony abruptly disappear. Loss of pollinators means a reduction in crop yields. Sublethal doses of insecticides (i.e. imidacloprid and other neonicotinoids) affect bee foraging behavior. However, research into the causes of CCD was inconclusive as of June 2007.

Hydroponics

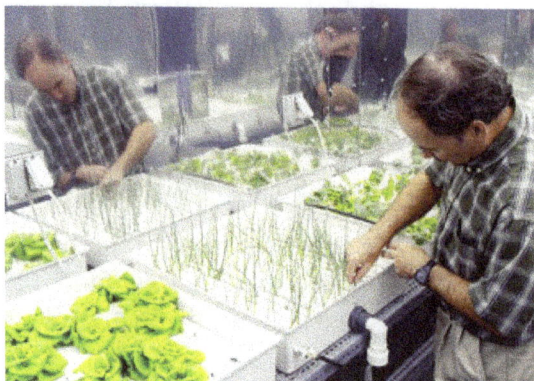

NASA researcher checking hydroponic onions with Bibb lettuce to his left and radishes to the right

Hydroponics is a subset of hydroculture, the method of growing plants without soil, using mineral nutrient solutions in a water solvent. Terrestrial plants may be grown with only their roots exposed to the mineral solution, or the roots may be supported by an inert medium, such as perlite or gravel. The nutrients in hydroponics can be from fish waste, duck manure, or *normal nutrients.*

History

The earliest published work on growing terrestrial plants without soil was the 1627 book *Sylva Sylvarum* by Francis Bacon, printed a year after his death. Water culture became a popular research technique after that. In 1699, John Woodward published his water culture experiments with spearmint. He found that plants in less-pure water sources grew better than plants in distilled water. By 1842, a list of nine elements believed to be essential for plant growth had been compiled, and the discoveries of German botanists Julius von Sachs and Wilhelm Knop, in the years 1859–1875, resulted in a development of the technique of soilless cultivation. Growth of terrestrial plants without soil in mineral nutrient solutions was called solution culture. It quickly became a standard research and teaching technique and is still widely used. Solution culture is now considered a type of hydroponics where there is no inert medium.

In 1929, William Frederick Gericke of the University of California at Berkeley began publicly promoting that solution culture be used for agricultural crop production. He first termed it aquaculture but later found that aquaculture was already applied to culture of aquatic organisms. Gericke created a sensation by growing tomato vines twenty-five feet high in his back yard in mineral nutrient solutions rather than soil. He introduced the term hydroponics, water culture, in 1937, proposed to him by W. A. Setchell, a phycologist with an extensive education in the classics.

Reports of Gericke's work and his claims that hydroponics would revolutionize plant agriculture prompted a huge number of requests for further information. Gericke had been denied use of the University's greenhouses for his experiments due to the administration's skepticism, and when the University tried to compel him to release his preliminary nutrient recipes developed at home he requested greenhouse space and time to improve them using appropriate research facilities. While he was eventually provided greenhouse space, the University assigned Hoagland and Arnon to re-develop Gericke's formula and show it held no benefit over soil grown plant yields, a view held by Hoagland. In 1940, Gericke published the book, *Complete Guide to Soil less Gardening,* after leaving his academic position in a climate that was politically unfavorable.

Two other plant nutritionists at the University of California were asked to research Gericke's claims. Dennis R. Hoagland and Daniel I. Arnon wrote a classic 1938 agricultural bulletin, *The Water Culture Method for Growing Plants Without Soil,*. Hoagland and Arnon claimed that hydroponic crop yields were no better than crop yields with good-quality soils. Crop yields were ultimately limited by factors other than mineral nutrients, especially light. This research, however, overlooked the fact that hydroponics has other advantages including the fact that the roots of the plant have constant access to oxygen and that the plants have access to as much or as little water as they need. This is important as one of the most common errors when growing is over- and under- watering; and hydroponics prevents this from occurring as large amounts of water can be made available to the plant and any water not used, drained away, recirculated, or actively aerated, eliminating anoxic conditions, which drown root systems in soil. In soil, a grower needs to be very experienced to know exactly how much water to feed the plant. Too much and the plant will be unable to access oxygen; too little and the plant will lose the ability to transport nutrients, which are typically moved into the roots while in solution. These two researchers developed several formulas for mineral nutrient solutions, known as Hoagland solution. Modified Hoagland solutions are still in use.

One of the earliest successes of hydroponics occurred on Wake Island, a rocky atoll in the Pacific Ocean used as a refuelling stop for Pan American Airlines. Hydroponics was used there in the 1930s to grow vegetables for the passengers. Hydroponics was a necessity on Wake Island because there was no soil, and it was prohibitively expensive to airlift in fresh vegetables.

In the 1960s, Allen Cooper of England developed the Nutrient film technique. The Land Pavilion at Walt Disney World's EPCOT Center opened in 1982 and prominently features a variety of hydroponic techniques. In recent decades, NASA has done extensive hydroponic research for its Controlled Ecological Life Support System (CELSS). Hydroponics intended to take place on Mars are using LED lighting to grow in a different color spectrum with much less heat.

Plants that are not traditionally grown in a climate would be possible to grow using a controlled environment system like hydroponics. NASA has also looked to utilize hydroponics in the space program. Ray Wheeler, a plant physiologist at Kennedy Space Center's Space Life Science Lab, believes that hydroponics will create advances within space travel. He terms this as a bioregenerative life support system.

Techniques

There are two main variations for each medium, sub-irrigation and top irrigation. For all techniques, most hydroponic reservoirs are now built of plastic, but other materials have been used including concrete, glass, metal, vegetable solids, and wood. The containers should exclude light to prevent algae growth in the nutrient solution.

Static Solution Culture

In static solution culture, plants are grown in containers of nutrient solution, such as glass Mason jars (typically, in-home applications), plastic buckets, tubs, or tanks. The solution is usually gently aerated but may be un-aerated. If un-aerated, the solution level is kept low enough that enough roots are above the solution so they get adequate oxygen. A hole is cut in the lid of the reservoir for each plant. There can be one to many plants per reservoir. Reservoir size can be increased as plant– size increases. A home made system can be constructed from plastic food containers or glass canning jars with aeration provided by an aquarium pump, aquarium airline tubing and aquarium valves. Clear containers are covered with aluminium foil, butcher paper, black plastic, or other material to exclude light, thus helping to eliminate the formation of algae. The nutrient solution is changed either on a schedule, such as once per week, or when the concentration drops below a certain level as determined with an electrical conductivity meter. Whenever the solution is depleted below a certain level, either water or fresh nutrient solution is added. A Mariotte's bottle, or a float valve, can be used to automatically maintain the solution level. In raft solution culture, plants are placed in a sheet of buoyant plastic that is floated on the surface of the nutrient solution. That way, the solution level never drops below the roots.

The deep water raft tank at the CDC South Aquaponics greenhouse in Brooks, Alberta.

Continuous-flow Solution Culture

The nutrient film technique being used to grow various salad greens

In continuous-flow solution culture, the nutrient solution constantly flows past the roots. It is much easier to automate than the static solution culture because sampling and adjustments to the temperature and nutrient concentrations can be made in a large storage tank that has potential to serve thousands of plants. A popular variation is the nutrient film technique or NFT, whereby a very shallow stream of water containing all the dissolved nutrients required for plant growth is recirculated past the bare roots of plants in a watertight thick root mat, which develops in the bottom of the channel and has an upper surface that, although moist, is in the air. Subsequent to this, an abundant supply of oxygen is provided to the roots of the plants. A properly designed NFT system is based on using the right channel slope, the right flow rate, and the right channel length. The main advantage of the NFT system over other forms of hydroponics is that the plant roots are exposed to adequate supplies of water, oxygen, and nutrients. In all other forms of production, there is a conflict between the supply of these requirements, since excessive or deficient amounts of one results in an imbalance of one or both of the others. NFT, because of its design, provides a system where all three requirements for healthy plant growth can be met at the same time, provided that the simple concept of NFT is always remembered and practised. The result of these advantages is that higher yields of high-quality produce are obtained over an extended period of cropping. A downside of NFT is that it has very little buffering against interruptions in the flow (e.g. power outages). But, overall, it is probably one of the more productive techniques.

The same design characteristics apply to all conventional NFT systems. While slopes along channels of 1:100 have been recommended, in practice it is difficult to build a base for channels that is sufficiently true to enable nutrient films to flow without ponding in locally depressed areas. As a consequence, it is recommended that slopes of 1:30 to 1:40 are used. This allows for minor irregularities in the surface, but, even with these slopes, ponding and water logging may occur. The slope may be provided by the floor, or benches or racks may hold the channels and provide the required slope. Both methods are used and depend on local requirements, often determined by the site and crop requirements.

As a general guide, flow rates for each gully should be 1 liter per minute. At planting, rates may be half this and the upper limit of 2 L/min appears about the maximum. Flow rates beyond these extremes are often associated with nutritional problems. Depressed growth rates of many crops have been observed when channels exceed 12 metres in length. On rapidly growing crops, tests have indicated that, while oxygen levels remain adequate, nitrogen may be depleted over the length of the gully. As a consequence, channel length should not exceed 10–15 metres. In situations where this is not possible, the reductions in growth can be eliminated by placing another nutrient feed halfway along the gully and halving the flow rates through each outlet.

Aeroponics

Aeroponics is a system wherein roots are continuously or discontinuously kept in an environment saturated with fine drops (a mist or aerosol) of nutrient solution. The method requires no substrate and entails growing plants with their roots suspended in a deep air or growth chamber with the roots periodically wetted with a fine mist of atomized nutrients. Excellent aeration is the main advantage of aeroponics.

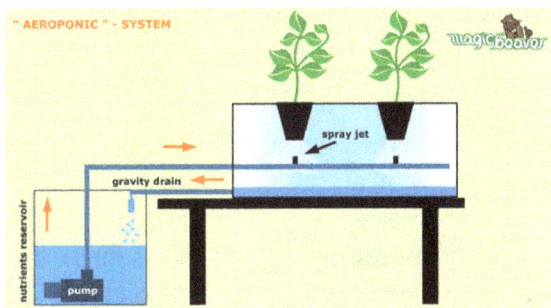

A diagram of the aeroponic technique.

Aeroponic techniques have proven to be commercially successful for propagation, seed germination, seed potato production, tomato production, leaf crops, and micro-greens. Since inventor Richard Stoner commercialized aeroponic technology in 1983, aeroponics has been implemented as an alternative to water intensive hydroponic systems worldwide. The limitation of hydroponics is the fact that 1 kilogram (2.2 lb) of water can only hold 8 milligrams (0.12 gr) of air, no matter whether aerators are utilized or not.

Another distinct advantage of aeroponics over hydroponics is that any species of plants can be grown in a true aeroponic system because the micro environment of an aeroponic can be finely controlled. The limitation of hydroponics is that only certain species of plants can survive for so long in water before they become waterlogged. The advantage of aeroponics is that suspended aeroponic plants receive 100% of the available oxygen and carbon dioxide to the roots zone, stems, and leaves, thus accelerating biomass growth and reducing rooting times. NASA research has shown that aeroponically grown plants have an 80% increase in dry weight biomass (essential minerals) compared to hydroponically grown plants. Aeroponics used 65% less water than hydroponics. NASA also concluded that aeroponically grown plants requires ¼ the nutrient input compared to hydroponics. Unlike hydroponically grown plants, aeroponically grown plants will not suffer transplant shock when transplanted to soil, and offers growers the ability to reduce the spread of disease and pathogens. Aeroponics is also widely used in laboratory studies of plant physiology and plant pathology. Aeroponic techniques have been given special attention from NASA since a mist is easier to handle than a liquid in a zero-gravity environment.

Fogponics

Fogponics is a derivation of aeroponics wherein the nutrient solution is aerosolized by a diaphragm vibrating at ultrasonic frequencies. Solution droplets produced by this method tend to be 5-10 μm in diameter, smaller than those produced by forcing a nutrient solution through pressurized nozzles, as in aeroponics. The smaller size of the droplets allows them to diffuse through the air more easily, and deliver nutrients to the roots without limiting their access to oxygen.

Passive Sub-irrigation

Passive sub-irrigation, also known as passive hydroponics or semi-hydroponics, is a method wherein plants are grown in an inert porous medium that transports water and fertilizer to the roots by capillary action from a separate reservoir as necessary, reducing labor and providing a constant supply of water to the roots. In the simplest method, the pot sits in a shallow solution of fertilizer and water or on a capillary mat saturated with nutrient solution. The various hydroponic media available, such as expanded clay and coconut husk, contain more air space than more traditional potting mixes, delivering increased oxygen to the roots, which is important in epiphytic plants such as orchids and bromeliads, whose roots are exposed to the air in nature. Additional advantages of passive hydroponics are the reduction of root rot and the additional ambient humidity provided through evaporations.

Ebb and Flow or Flood and Drain Sub-irrigation

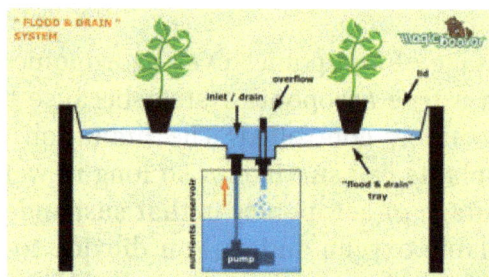

A Ebb and flow or flood and drain hydroponics system.

In its simplest form, there is a tray above a reservoir of nutrient solution. Either the tray is filled with growing medium (clay granules being the most common) and planted directly or pots of medium stand in the tray. At regular intervals, a simple timer causes a pump to fill the upper tray with nutrient solution, after which the solution drains back down into the reservoir. This keeps the medium regularly flushed with nutrients and air. Once the upper tray fills past the drain stop, it begins recirculating the water until the timer turns the pump off, and the water in the upper tray drains back into the reservoirs.

Run to Waste

In a run-to-waste system, nutrient and water solution is periodically applied to the medium surface. The method was invented in Bengal in 1946, for this reason it is sometimes referred to as "The Bengal System".

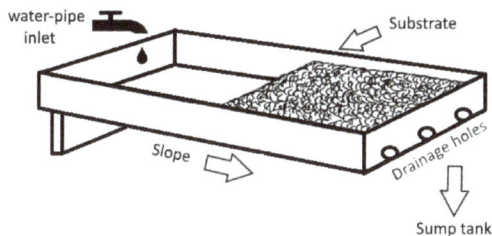

A run-to-waste hydroponics system referred to as "The Bengal System" after the region in northeastern India where it was invented (circa 1946–1948).

This method can be setup in various configurations. In its simplest form, a nutrient-and-water solution is manually applied one or more times per day to a container of inert growing media, such as rockwool, perlite, vermiculite, coco fibre, or sand. In a slightly more complex system, it is automated with a delivery pump, a timer and irrigation tubing to deliver nutrient solution with a delivery frequency that is governed by the key parameters of plant size, plant growing stage, climate, substrate, and substrate conductivity, pH, and water content.

In a commercial setting, watering frequency is multi-factorial and governed by computers or PLCs.

Commercial hydroponics production of large plants like tomatoes, cucumber, and peppers use one form or another of run-to-waste hydroponics.

In environmentally responsible uses, the nutrient rich waste is collected and processed through an on site filtration system to be used many times, making the system very productive.

The majority of bonsai are now grown in soil-free substrates (typically consisting of akadama, grit, diatomaceous earth and other inorganic components) and have their water and nutrients provided in a run-to-waste form.

Deep Water Culture

The Deep water culture technique being used to grow Hungarian wax peppers.

The hydroponic method of plant production by means of suspending the plant roots in a solution of nutrient-rich, oxygenated water. Traditional methods favor the use of plastic buckets and large containers with the plant contained in a net pot suspended from the centre of the lid and the roots suspended in the nutrient solution. The solution is oxygen saturated by an air pump combined with porous stones. With this method, the plants grow much faster because of the high amount of oxygen that the roots receive.

Top-fed Deep Water Culture

Top-fed deep water culture is a technique involving delivering highly oxygenated nutrient solution direct to the root zone of plants. While deep water culture involves the plant roots hanging down into a reservoir of nutrient solution, in top-fed deep water culture the solution is pumped from the reservoir up to the roots (top feeding). The water is released over the plant's roots and then runs back into the reservoir below in a constantly recirculating system. As with deep water culture, there is an airstone in the reservoir that pumps air into the water via a hose from outside the reservoir. The airstone helps add oxygen to the water. Both the airstone and the water pump run 24 hours a day.

The biggest advantage of top-fed deep water culture over standard deep water culture is increased growth during the first few weeks. With deep water culture, there is a time when the roots have not reached the water yet. With top-fed deep water culture, the roots get easy access to water from the beginning and will grow to the reservoir below much more quickly than with a deep water culture system. Once the roots have reached the reservoir below, there is not a huge advantage with top-fed deep water culture over standard deep water culture. However, due to the quicker growth in the beginning, grow time can be reduced by a few weeks.

Rotary

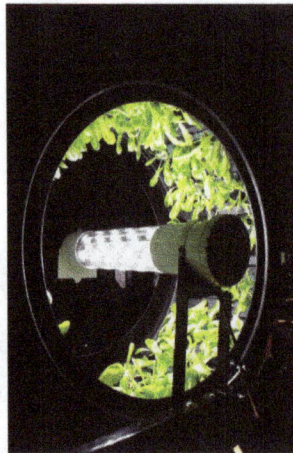

A Rotary hydroponic cultivation demonstration at the Belgian Pavilion Expo in 2015.

A rotary hydroponic garden is a style of commercial hydroponics created within a circular frame which rotates continuously during the entire growth cycle of whatever plant is being grown.

While system specifics vary, systems typically rotate once per hour, giving a plant 24 full turns within the circle each 24-hour period. Within the center of each rotary hydroponic garden is a high intensity grow light, designed to simulate sunlight, often with the assistance of a mechanized timer.

Each day, as the plants rotate, they are periodically watered with a hydroponic growth solution to provide all nutrients necessary for robust growth. Due to the plants continuous fight against gravity, plants typically mature much more quickly than when grown in soil or other traditional hydroponic growing systems. Due to the small foot print a rotary hydroponic system has, it allows for more plant material to be grown per sq foot of floor space than other traditional hydroponic systems.

Substrates

One of the most obvious decisions hydroponic farmers have to make is which medium they should use. Different media are appropriate for different growing techniques.

Expanded Clay Aggregate

Expanded clay pebbles.

Baked clay pellets are suitable for hydroponic systems in which all nutrients are carefully controlled in water solution. The clay pellets are inert, pH neutral and do not contain any nutrient value.

The clay is formed into round pellets and fired in rotary kilns at 1,200 °C (2,190 °F). This causes the clay to expand, like popcorn, and become porous. It is light in weight, and does not compact over time. The shape of an individual pellet can be irregular or uniform depending on brand and manufacturing process. The manufacturers consider expanded clay to be an ecologically sustainable and re-usable growing medium because of its ability to be cleaned and sterilized, typically by washing in solutions of white vinegar, chlorine bleach, or hydrogen peroxide (H_2O_2), and rinsing completely.

Another view is that clay pebbles are best not re-used even when they are cleaned, due to root growth that may enter the medium. Breaking open a clay pebble after a crop has been grown will reveal this growth.

Growstones

Growstones, made from glass waste, have both more air and water retention space than perlite and peat. This aggregate holds more water than parboiled rice hulls. Growstones by volume consist of 0.5 to 5% calcium carbonate – for a standard 5.1 kg bag of Growstones that corresponds to 25.8 to 258 grams of calcium carbonate. The remainder is soda-lime glass.

Coir Peat

Coco peat, also known as coir or coco, is the leftover material after the fibres have been removed from the outermost shell (bolster) of the coconut. Coir is a 100% natural grow and flowering medium. Coconut coir is colonized with trichoderma fungi, which protects roots and stimulates root growth. It is extremely difficult to over-water coir due to its perfect air-to-water ratio; plant roots thrive in this environment. Coir has a high cation exchange, meaning it can store unused minerals to be released to the plant as and when it requires it. Coir is available in many forms; most common is coco peat, which has the appearance and texture of soil but contains no mineral content.

Rice Husks

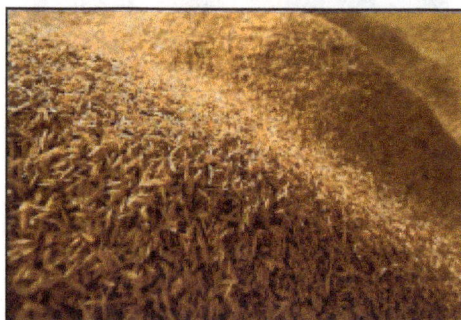

Rice husks, a hydroponic growing substrate option.

Parboiled rice husks (PBH) are an agricultural byproduct that would otherwise have little use. They decay over time, and allow drainage, and even retain less water than growstones. A study showed that rice husks did not affect the effects of plant growth regulators.

Perlite

Perlite, a hydroponic growing substrate option.

Perlite is a volcanic rock that has been superheated into very lightweight expanded glass pebbles. It is used loose or in plastic sleeves immersed in the water. It is also used in potting soil mixes to decrease soil density. Perlite has similar properties and uses to vermiculite but, in general, holds more air and less water. If not contained, it can float if flood and drain feeding is used. It is a fusion of granite, obsidian, pumice and basalt. This volcanic rock is naturally fused at high temperatures undergoing what is called "Fusionic Metamorphosis".

Vermiculite

Vermiculite close-up.

Like perlite, vermiculite is a mineral that has been superheated until it has expanded into light pebbles. Vermiculite holds more water than perlite and has a natural "wicking" property that can draw water and nutrients in a passive hydroponic system. If too much water and not enough air surrounds the plants roots, it is possible to gradually lower the medium's water-retention capability by mixing in increasing quantities of perlite.

Pumice

A pumice stone.

Like perlite, pumice is a lightweight, mined volcanic rock that finds application in hydroponics.

Sand

Sand is cheap and easily available. However, it is heavy, does not hold water very well, and it must be sterilized between uses.

Gravel

The same type that is used in aquariums, though any small gravel can be used, provided it is washed first. Indeed, plants growing in a typical traditional gravel filter bed, with water circulated using electric powerhead pumps, are in effect being grown using gravel hydroponics. Gravel is inexpensive, easy to keep clean, drains well and will not become waterlogged. However, it is also heavy, and, if the system does not provide continuous water, the plant roots may dry out.

Wood Fibre

Excelsior, or wood wool

Wood fibre, produced from steam friction of wood, is a very efficient organic substrate for hydroponics. It has the advantage that it keeps its structure for a very long time. Wood wool (i.e. wood slivers) have been used since the earliest days of the hydroponics research. However, more recent research suggests that wood fibre may have detrimental effects on "plant growth regulators".

Sheep Wool

Wool from shearing sheep is a little-used yet promising renewable growing medium. In a study comparing wool with peat slabs, coconut fibre slabs, perlite and rockwool slabs to grow cucumber plants, sheep wool had a greater air capacity of 70%, which decreased with use to a comparable 43%, and water capacity that increased from 23% to 44% with use. Using sheep wool resulted in the greatest yield out of the tested substrates, while application of a biostimulator consisting of humic acid, lactic acid and Bacillus subtilis improved yields in all substrates.

Rock Wool

Rock wool (mineral wool) is the most widely used medium in hydroponics. Rock wool is an inert substrate suitable for both run-to-waste and recirculating systems. Rock wool is made from molten rock, basalt or 'slag' that is spun into bundles of single filament fibres, and bonded into a medium capable of capillary action, and is, in effect, protected from most common microbiological degradation. Rock wool has many advantages and some disadvantages. The latter being the possible skin irritancy (mechanical) whilst handling (1:1000). Flushing with cold water usually brings relief. Advantages include its proven efficiency and effectiveness as a commercial hydroponic substrate. Most of the rock wool sold to date is a non-hazardous, non-carcinogenic material, falling under Note Q of the European Union Classification Packaging and Labeling Regulation (CLP).

Rock wool close-up.

Brick Shards

Brick shards have similar properties to gravel. They have the added disadvantages of possibly altering the pH and requiring extra cleaning before reuse.

Polystyrene Packing Peanuts

Polystyrene foam peanuts

Polystyrene packing peanuts are inexpensive, readily available, and have excellent drainage. However, they can be too lightweight for some uses. They are used mainly in closed-tube systems. Note that polystyrene peanuts must be used; biodegradable packing peanuts will decompose into a sludge. Plants may absorb styrene and pass it to their consumers; this is a possible health risk.

Nutrient Solutions

Inorganic Hydroponic Solutions

The formulation of hydroponic solutions is an application of plant nutrition, with nutrient deficiency symptoms mirroring those found in traditional soil based agriculture. However, the underlying chemistry of hydroponic solutions can differ from soil chemistry in many significant ways. Important differences include:

- Unlike soil, hydroponic nutrient solutions do not have cation-exchange capacity (CEC) from clay particles or organic matter. The absence of CEC means the pH and nutrient concentrations can change much more rapidly in hydroponic setups than is possible in soil.

- Selective absorption of nutrients by plants often imbalances the amount of counterions in solution. This imbalance can rapidly affect solution pH and the ability of plants to absorb nutrients of similar ionic charge. For instance, nitrate anions are often consumed rapidly by plants to form proteins, leaving an excess of cations in solution. This cation imbalance can lead to deficiency symptoms in other cation based nutrients (e.g. Mg^{2+}) even when an ideal quantity of those nutrients are dissolved in the solution.

- Depending the on pH, and/or the presence of water contaminants, nutrients, such as iron, can precipitate from the solution and become unavailable to plants. Routine adjustments to pH, buffering the solution, and/or the use of chelating agents is often necessary.

As in conventional agriculture, nutrients should be adjusted to satisfy Liebig's law of the minimum for each specific plant variety. Nevertheless, generally acceptable concentrations for nutrient solutions exist, with minimum and maximum concentration ranges for most plants being somewhat similar. Most nutrient solutions are mixed to have concentrations between 1,000 and 2,500 ppm. Acceptable concentrations for the individual nutrient ions, which comprise that total ppm figure, are summarized in the following table. For essential nutrients, concentrations below these ranges often lead to nutrient deficiencies while exceeding these ranges can lead to nutrient toxicity. Optimum nutrition concentrations for plant varieties are found empirically by experience and/or by plant tissue tests.

Organic Hydroponic Solutions

Organic fertilizers can be used to supplement or entirely replace the inorganic compounds used in conventional hydroponic solutions. However, using organic fertilizers introduces a number of challenges that are not easily resolved. Examples include:

- organic fertilizers are highly variable in their nutritional compositions. Even similar materials can differ significantly based on their source (e.g. the quality of manure varies based on an animal's diet).

- organic fertilizers are often sourced from animal byproducts, making disease transmission a serious concern for plants grown for human consumption or animal forage.

- organic fertilizers are often particulate and can clog substrates or other growing equipment. Sieving and/or milling the organic materials to fine dusts is often necessary.

- some organic materials (i.e. particularly manures and offal) can further degrade to emit foul odors.

Nevertheless, if precautions are taken, organic fertilizers can be used successfully in hydroponics.

Organically Sourced Macronutrients

Examples of suitable materials, with their average nutritional contents tabulated in terms of percent dried mass, are listed in the following table.

Organically Sourced Micronutrients

Micronutrients can be sourced from organic fertilizers as well. For example, composted pine bark is high in manganese and is sometimes used to fulfill that mineral requirement in hydroponic solutions. To satisfy requirements for National Organic Programs, pulverized, unrefined minerals (e.g. Gypsum, Calcite, and glauconite) can also be added to satisfy a plant's nutritional needs.

Additives

In addition to chelating agents, humic acids can be added to increase nutrient uptake.

Tools

Common Equipment

Managing nutrient concentrations and pH values within acceptable ranges is essential for successful hydroponic horticulture. Common tools used to manage hydroponic solutions include:

- Electrical conductivity meters, a tool which estimates nutrient ppm by measuring how well a solution transmits an electric current.

- pH meter, a tool that uses an electric current to determine the concentration of hydrogen ions in solution.

- Litmus paper, disposable pH indicator strips that determine hydrogen ion concentrations by color changing chemical reaction.

- Graduated cylinders or measuring spoons to measure out premixed, commercial hydroponic solutions.

Advanced Equipment

Advanced equipment can also be used to perform accurate chemical analyses of nutrient solutions. Examples include:

- Balances for accurately measuring materials.

- Laboratory glassware, such as burettes and pipettes, for performing titrations.

- Colorimeters for solution tests which apply the Beer–Lambert law.

Using advanced equipment for hydroponic solutions can be beneficial to growers of any background because nutrient solutions are often reusable. Because nutrient solutions are virtually never completely depleted, and should never be due to the unacceptably low osmotic pressure that would result, re-fortification of old solutions with new nutrients can save growers money and can control point source pollution, a common source for the eutrophication of nearby lakes and streams.

Software

Although pre-mixed concentrated nutrient solutions are generally purchased from commercial nutrient manufacturers by hydroponic hobbyists and small commercial growers, several tools exist to help anyone prepare their own solutions without extensive knowledge about chemistry. The free and open source tools HydroBuddy and HydroCal have been created by professional chemists to help any hydroponics grower prepare their own nutrient solutions. The first program is available for Windows, Mac and Linux while the second one can be used through a simple JavaScript interface. Both programs allow for basic nutrient solution preparation although HydroBuddy provides added functionality to use and save custom substances, save formulations and predict electrical conductivity values.

Mixing Solutions

Often mixing hydroponic solutions using individual salts is impractical for hobbyists and/or small-scale commercial growers because commercial products are available at reasonable prices. However, even when buying commercial products, multi-component fertilizers are popular. Often these products are bought as three part formulas which emphasize certain nutritional roles. For example, solutions for vegetative growth (i.e. high in nitrogen), flowering (i.e. high in potassium and phosphorus), and micronutrient solutions (i.e. with trace minerals) are popular. The timing and application of these multi-part fertilizers should coincide with a plant's growth stage. For example, at the end of an annual plant's life cycle, a plant should be restricted from high nitrogen fertilizers. In most plants, nitrogen restriction inhibits vegetative growth and helps induce flowering.

Advancements

With pest problems reduced and nutrients constantly fed to the roots, productivity in hydroponics is high; however, growers can further increase yield by manipulating a plant's environment by constructing sophisticated growrooms.

CO$_2$ Enrichment

To increase yield further, some sealed greenhouses inject CO$_2$ into their environment to help improve growth and plant fertility.

Fungicide

Fungicides are biocidal chemical compounds or biological organisms used to kill fungi or fungal spores. A fungistatic inhibits their growth. Fungi can cause serious damage in agriculture, resulting in critical losses of yield, quality, and profit. Fungicides are used both in agriculture and to fight fungal infections in animals. Chemicals used to control oomycetes, which are not fungi, are also referred to as fungicides, as oomycetes use the same mechanisms as fungi to infect plants.

Fungicides can either be contact, translaminar or systemic. Contact fungicides are not taken up into the plant tissue and protect only the plant where the spray is deposited. Translaminar fungicides redistribute the fungicide from the upper, sprayed leaf surface to the lower, unsprayed surface. Systemic fungicides are taken up and redistributed through the xylem vessels. Few fungicides move to all parts of a plant. Some are locally systemic, and some move upwardly.

Most fungicides that can be bought retail are sold in a liquid form. A very common active ingredient is sulfur, present at 0.08% in weaker concentrates, and as high as 0.5% for more potent fungicides. Fungicides in powdered form are usually around 90% sulfur and are very toxic. Other active ingredients in fungicides include neem oil, rosemary oil, jojoba oil, the bacterium *Bacillus subtilis*, and the beneficial fungus *Ulocladium oudemansii*.

Fungicide residues have been found on food for human consumption, mostly from post-harvest treatments. Some fungicides are dangerous to human health, such as vinclozolin, which has now been removed from use. Ziram is also a fungicide that is thought to be toxic to humans if exposed to chronically. A number of fungicides are also used in human health care.

Natural Fungicides

Plants and other organisms have chemical defenses that give them an advantage against microorganisms such as fungi. Some of these compounds can be used as fungicides:

- Tea tree oil
- Cinnamaldehyde
- Citronella oil
- Jojoba oil
- Nimbin
- Oregano oil
- Rosemary oil
- Monocerin
- Milk

Whole live or dead organisms that are efficient at killing or inhibiting fungi can sometimes be used as fungicides:

- *Bacillus subtilis*
- *Ulocladium oudemansii*
- Kelp (powdered dried kelp is fed to cattle to help prevent fungal infection)
- *Ampelomyces quisqualis*

Resistance

Pathogens respond to the use of fungicides by evolving resistance. In the field several mechanisms of resistance have been identified. The evolution of fungicide resistance can be gradual or sudden. In qualitative or discrete resistance, a mutation (normally to a single gene) produces a race of a fungus with a high degree of resistance. Such resistant varieties also tend to show stability, persisting after the fungicide has been removed from the market. For example, sugar beet leaf blotch remains resistant to azoles years after they were no longer used for control of the disease. This is because such mutations often have a high selection pressure when the fungicide is used, but there is low selection pressure to remove them in the absence of the fungicide.

In instances where resistance occurs more gradually, a shift in sensitivity in the pathogen to the fungicide can be seen. Such resistance is polygenic – an accumulation of many mutations in different genes, each having a small additive effect. This type of resistance is known as quantitative or continuous resistance. In this kind of resistance, the pathogen population will revert to a sensitive state if the fungicide is no longer applied.

Little is known about how variations in fungicide treatment affect the selection pressure to evolve resistance to that fungicide. Evidence shows that the doses that provide the most control of the disease also provide the largest selection pressure to acquire resistance, and that lower doses decrease the selection pressure.

In some cases when a pathogen evolves resistance to one fungicide, it automatically obtains resistance to others – a phenomenon known as cross resistance. These additional fungicides are normally of the same chemical family or have the same mode of action, or can be detoxified by the same mechanism. Sometimes negative cross resistance occurs, where resistance to one chemical class of fungicides leads to an increase in sensitivity to a different chemical class of fungicides. This has been seen with carbendazim and diethofencarb.

There are also recorded incidences of the evolution of multiple drug resistance by pathogens – resistance to two chemically different fungicides by separate mutation events. For example, *Botrytis cinerea* is resistant to both azoles and dicarboximide fungicides.

There are several routes by which pathogens can evolve fungicide resistance. The most common mechanism appears to be alteration of the target site, in particular as a defence against single site of action fungicides. For example, Black Sigatoka, an economically important pathogen of banana, is resistant to the QoI fungicides, due to a single nucleotide change resulting in the replacement of one amino acid (glycine) by another (alanine) in the target protein of the QoI fungicides, cytochrome b. It is presumed that this disrupts the binding of the fungicide to the protein, rendering the fungicide ineffective. Upregulation of target genes can also render the fungicide ineffective. This is seen in DMI-resistant strains of *Venturia inaequalis*.

Resistance to fungicides can also be developed by efficient efflux of the fungicide out of the cell. *Septoria tritici* has developed multiple drug resistance using this mechanism. The pathogen had 5 ABC-type transporters with overlapping substrate specificities that together work to pump toxic chemicals out of the cell.

In addition to the mechanisms outlined above, fungi may also develop metabolic pathways that circumvent the target protein, or acquire enzymes that enable metabolism of the fungicide to a harmless substance.

Fungicide Resistance Management

The fungicide resistance action committee (FRAC) has several recommended practices to try to avoid the development of fungicide resistance, especially in at-risk fungicides including *Strobilurins* such as azoxystrobin.

Products should not be used in isolation, but rather as mixture, or alternate sprays, with another fungicide with a different mechanism of action. The likelihood of the pathogen's developing resistance is greatly decreased by the fact that any resistant isolates to one fungicide will be killed by the other; in other words, two mutations would be required rather than just one. The effectiveness of this technique can be demonstrated by Metalaxyl, a phenylamide fungicide. When used as the sole product in Ireland to control potato blight (*Phytophthora infestans*), resistance developed within one growing season. However, in countries like the UK where it was marketed only as a mixture, resistance problems developed more slowly.

Fungicides should be applied only when absolutely necessary, especially if they are in an at-risk group. Lowering the amount of fungicide in the environment lowers the selection pressure for resistance to develop.

Manufacturers' doses should always be followed. These doses are normally designed to give the right balance between controlling the disease and limiting the risk of resistance development. Higher doses increase the selection pressure for single-site mutations that confer resistance, as all strains but those that carry the mutation will be eliminated, and thus the resistant strain will propagate. Lower doses greatly increase the risk of polygenic resistance, as strains that are slightly less sensitive to the fungicide may survive.

It is also recommended that where possible fungicides are used only in a protective manner, rather than to try to cure already-infected crops. Far fewer fungicides have curative/eradicative ability than protectant. Thus, fungicide preparations advertised as having curative action may have only one active chemical; a single fungicide acting in isolation increases the risk of fungicide resistance.

It is better to use an integrative pest management approach to disease control rather than relying on fungicides alone. This involves the use of resistant varieties and hygienic practices, such as the removal of potato discard piles and stubble on which the pathogen can overwinter, greatly reducing the titre of the pathogen and thus the risk of fungicide resistance development.

Bromacil

Bromacil is an organic compound with the chemical formula $C_9H_{13}BrN_2O_2$, commercially available as a herbicide. Bromacil was first registered as a pesticide in the U.S. in 1961. It is used for brush control and non-cropland areas. It works by interfering with photosynthesis by entering the plant through the root zone and moving throughout the plant. Bromacil is one of a group of compounds called substituted uracils. These materials are broad spectrum herbicides used for nonselective weed and brush control on non-croplands, as well as for selective weed control on a limited number of crops, such as citrus fruit and pineapple. Bromacil is also found to be excellent at controlling perennial grasses.

Safety

There are quite a few safety precautions that should be taken when dealing with Bromacil. Dry formulations containing bromacil must bear the word "Caution" and liquid formulas must signal "Warning." Care should be exercised when spraying Bromacil on plants because it will also stop the photosynthesis of the adjacent non-target plants, therefore killing them. Bromacil should never be used in residential or recreation areas for risk of exposure. Bromacil is slightly toxic if individuals accidentally eat or touch residues and practically nontoxic if inhaled. Bromacil is a mild eye irritant and a very slight skin irritant. It is not a skin sensitizer. In studies using laboratory animals, bromacil is slightly toxic by the oral, dermal, and inhalation routes and has been placed in Toxicity Category IV (the lowest of four categories) for these effects. This herbicide should be stored in a cool, dry place and after any handling a thorough hand-washing is advised.

In regards to occupational exposure, the National Institute for Occupational Safety and Health has recommended workers handing bromacil not exceed an exposure of 1 ppm (10 mg/m^3) over an eight-hour time-weighted average.

Facts

Bromacil (40%) is combined with the active ingredient diuron in the herbicide Krovar, which is used by companies such as the Washington State Department of Transportation (WSDOT). It is in a group of chemicals that are absorbed through the gut and excreted primarily in the urine. The half-life of bromacil in soils is about 60 days, but as long as 8 months in some conditions. Bromacil is available in granular, liquid, water-soluble liquid, and wettable powder formulations. Because bromacil is a possible human carcinogen and systemic toxicity may result from intermediate exposure (one week to several months), U.S. Environmental Protection Agency (EPA) assessed risk to workers using several major exposure scenarios. Bromacil is stable to hydrolysis under normal environmental conditions.

Applications

Bromacil is applied mainly by sprayers including boom, hand-held, knapsack, compressed air, tank-type, and power sprayers. Bromacil is also applied using aerosol, shaker, or sprinkler cans. Solid forms of bromacil are spread using granule applicators and spreaders. Application using aircraft is allowed only for Special Local Need registrations to control vegetation on the Department of Defense's Yakima Firing Center in the state of Washington.

Bentazon

Bentazon (Bentazone, Basagran, Herbatox, Leader, Laddock) is a chemical manufactured by BASF Chemicals for use in herbicides. It is categorized under the thiadiazine group of chemicals. Sodium bentazon is available commercially and appears slightly brown in colour.

Usage

Bentazon is a selective herbicide as it only damages plants unable to metabolize the chemical. It is considered safe for use on alfalfa, beans (with the exception of garbanzo beans), corn, peanuts, peas (with the exception of blackeyed peas), pepper, peppermint, rice, sorghum, soybeans and spearmint; as well as lawns and turf. Bentazon is usually applied aerially or through contact spraying on food crops to control the spread of weeds occurring amongst food crops. Herbicides containing bentazon should be kept away from high heat as it will release toxic sulfur and nitrogen fumes.

Bentazon is currently registered for use in the United States in accordance with requirements set forth by the United States Environmental Protection Agency. However as of September 2010, the herbicides Basagran M60, Basagran DF, Basagran AG, Prompt 5L and Laddock 5L are currently under review for pending requests for voluntary registration cancellation.

Water and Ground Contamination

In general, bentazon is quickly metabolized and degraded by both plants and animals. However, soil leaching and runoff is a major concern in terms of water contamination. In 1995 the Environmental Protection Agency (EPA) stated that levels of bentazon in both ground water and surface water "exceed levels of concern". Despite the establishment of a 20 parts per billion Health Advisory Level there is no requirement to measure for bentazon in water supplies as the Safe Drinking Water Act does not regulate bentazon. The United States EPA found bentazon in 64 out of 200 wells in California - the highest number of detections in their 1995 study. This prompted the State of California to review existing toxicology studies and establish a "Public Health Goal" that limits bentazon in drinking water to 200 parts per billion.

The EPA requires ground water and environmental hazard advisory labels on all commercially available herbicides containing bentazon. Both statements warn against application and/or disposal of bentazon directly into water, or in areas where soil leaching is common.

Food Contamination

A number of limits have been placed on bentazon to reduce the possibility of toxic effects on humans. Tolerance levels vary depending on the use of the food/animal product. The following tolerance levels for bentazon have been established in the United States:

- 0.02 ppm for milk.

- 0.05 ppm (parts per million) for meat and animal byproducts (poultry, eggs, cattle, hogs, sheep and goats).

- 0.05 ppm for dried beans (excluding soybeans), corn (fresh and grain), bohemian chili peppers, peanuts, rice, soybeans, and sorghum used for fodder and grain.

- 0.5 ppm for succulent beans and peas.

- 0.3 ppm for peanut hulls.

- 1 ppm for mint and dried peas.

- 3 ppm for rice (straw), corn for fodder and forage, and peanuts used in hay and forage.

- 8 ppm for pea vine hays (dried), and soybeans used for foraging or hay.

It is recommended that food and feed supplies be stored away from herbicides containing bentazon. Aerial spraying should be conducted in a manner that prevents spray drift towards water sources and food crops susceptible to bentazon.

Toxicity to Nonhuman Species

A 1994 study concluded that bentazon is non-toxic to honeybees, and is not harmful to beetles. Studies have found that bentazon is toxic to rainbow trout and bluegill sunfish at 190 ppm and 616 ppm, respectively. Bentazon is considered toxic to birds as it affects their reproductive capacities.

Among mammals, bentazon is found to be moderately toxic when ingested or absorbed through the skin. Lethal doses (LD_{50}, the dose required to kill half the population being studied) for bentazon have been established for:

- Cats: 500 mg/kg

- Rats: 1100 mg/kg to 2063 mg/kg

- Mice: 400 mg/kg

- Rabbits: 750 mg/kg

Dogs in a study being fed 13.1 mg of bentazon a day developed diarrhoea, anemia and dehydration. In another study using dogs, prostate inflammation was also observed along with previously noted health effects In experiments conducted on hamsters, mice and rats, bentazon was not found to cause gene mutations to damage to DNA and chromosomes.

Toxicity to Humans

Bentazon has been classified by the EPA as a "Group E" chemical, because it is believed to be non-carcinogenic to humans (as based on testing conducted on animals). However, there are no studies or experiments that can determine toxic and/or carcinogenic effects of bentazon on humans. Workers applying the herbicide would be most exposed to bentazon, and so have been advised to wear protective clothing (goggles, gloves and aprons) at all times when handling the chemical. Bentazon causes allergy-like symptoms as it irritates the eyes, skin and respiratory tract. Ingesting bentazon causes nausea, diarrhoea, trembling, vomiting and difficulty breathing. Workers handling bentazon must wash their hands before eating, drinking, smoking, and using the bathroom to minimize contact with skin. The effects of bentazon ingestion has been observed in humans who chose the herbicide to commit suicide. Ingestion of bentazon was observed to cause fevers, renal failure (kidney failure), accelerated heart rate (tachycardia), shortness of breath (dyspnea) and hyperthermia. Ingestion of 88 grams of bentazon caused death in an adult.

Herbicide

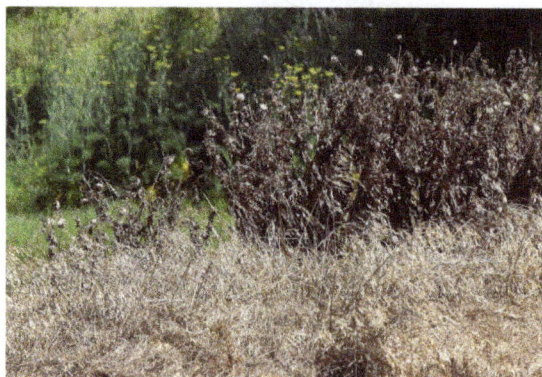

Weeds controlled with herbicide

Herbicide(s), also commonly known as weedkillers, are chemical substances used to control unwanted plants. Selective herbicides control specific weed species, while leaving the desired crop relatively unharmed, while non-selective herbicides (sometimes called "total weedkillers" in commercial products) can be used to clear waste ground, industrial and construction sites, railways and railway embankments as they kill all plant material with which they come into contact. Apart from selective/non-selective, other important distinctions include *persistence* (also known as *residual action*: how long the product stays in place and remains active), *means of uptake* (whether it is absorbed by above-ground foliage only, through the roots, or by other means), and *mechanism of action* (how it works). Historically, products such as common salt and other metal salts were used as herbicides, however these have gradually fallen out of favor and in some countries a number of these are banned due to their persistence in soil, and toxicity and groundwater contamination concerns. Herbicides have also been used in warfare and conflict.

Modern herbicides are often synthetic mimics of natural plant hormones which interfere with growth of the target plants. The term organic herbicide has come to mean herbicides intended for organic farming; these are often less efficient and more costly than synthetic herbicides and are based on natural materials. Some plants also produce their own natural herbicides, such as the genus *Juglans* (walnuts), or the tree of heaven; such action of natural herbicides, and other related chemical interactions, is called allelopathy. Due to herbicide resistance - a major concern in agriculture - a number of products also combine herbicides with different means of action.

In the US in 2007, about 83% of all herbicide usage, determined by weight applied, was in agriculture. In 2007, world pesticide expenditures totaled about \$39.4 billion; herbicides were about 40% of those sales and constituted the biggest portion, followed by insecticides, fungicides, and other types. Smaller quantities are used in forestry, pasture systems, and management of areas set aside as wildlife habitat.

History

Prior to the widespread use of chemical herbicides, cultural controls, such as altering soil pH, salinity, or fertility levels, were used to control weeds. Mechanical control (including tillage) was also (and still is) used to control weeds.

First Herbicides

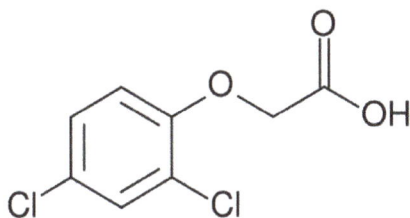

2,4-D, the first chemical herbicide, was discovered during the Second World War.

Although research into chemical herbicides began in the early 20th century, the first major breakthrough was the result of research conducted in both the UK and the US during the Second World War into the potential use of agents as biological weapons. The first modern herbicide, 2,4-D, was first discovered and synthesized by W. G. Templeman at Imperial Chemical Industries. In 1940, he showed that "Growth substances applied appropriately would kill certain broad-leaved weeds in cereals without harming the crops." By 1941, his team succeeded in synthesizing the chemical. In the same year, Pokorny in the US achieved this as well.

Independently, a team under Juda Hirsch Quastel, working at the Rothamsted Experimental Station made the same discovery. Quastel was tasked by the Agricultural Research Council (ARC) to discover methods for improving crop yield. By analyzing soil as a dynamic system, rather than an inert substance, he was able to apply techniques such as perfusion. Quastel was able to quantify the influence of various plant hormones, inhibitors and other chemicals on the activity of microorganisms in the soil and assess their direct impact on plant growth. While the full work of the unit remained secret, certain discoveries were developed for commercial use after the war, including the 2,4-D compound.

When it was commercially released in 1946, it triggered a worldwide revolution in agricultural output and became the first successful selective herbicide. It allowed for greatly enhanced weed control in wheat, maize (corn), rice, and similar cereal grass crops, because it kills dicots (broadleaf plants), but not most monocots (grasses). The low cost of 2,4-D has led to continued usage today, and it remains one of the most commonly used herbicides in the world. Like other acid herbicides, current formulations use either an amine salt (often trimethylamine) or one of many esters of the parent compound. These are easier to handle than the acid.

Further Discoveries

The triazine family of herbicides, which includes atrazine, were introduced in the 1950s; they have the current distinction of being the herbicide family of greatest concern regarding groundwater contamination. Atrazine does not break down readily (within a few weeks) after being applied to soils of above neutral pH. Under alkaline soil conditions, atrazine may be carried into the soil profile as far as the water table by soil water following rainfall causing the aforementioned contamination. Atrazine is thus said to have "carryover", a generally undesirable property for herbicides.

Glyphosate (Roundup) was introduced in 1974 for nonselective weed control. Following the development of glyphosate-resistant crop plants, it is now used very extensively for selective weed control in growing crops. The pairing of the herbicide with the resistant seed contributed to the consolidation of the seed and chemistry industry in the late 1990s.

Many modern chemical herbicides used in agriculture and gardening are specifically formulated to decompose within a short period after application. This is desirable, as it allows crops and plants to be planted afterwards, which could otherwise be affected by the herbicide. However, herbicides with low residual activity (i.e., that decompose quickly) often do not provide season-long weed control and do not ensure that weed roots are killed beneath construction and paving (and cannot emerge destructively in years to come), therefore there remains a role for weedkiller with high levels of persistence in the soil.

Terminology

Herbicides are classified/grouped in various ways e.g. according to the activity, timing of application, method of application, mechanism of action, chemical family. This gives rise to a considerable level of terminology related to herbicides and their use.

Intended Outcome

- Control is the destruction of unwanted weeds, or the damage of them to the point where they are no longer competitive with the crop.

- Suppression is incomplete control still providing some economic benefit, such as reduced competition with the crop.

- Crop safety, for selective herbicides, is the relative absence of damage or stress to the crop. Most selective herbicides cause some visible stress to crop plants.

- Defoliant, similar to herbicides, but designed to remove foliage (leaves) rather than kill the plant.

Selectivity (All Plants or Specific Plants)

- Selective herbicides: They control or suppress certain plants without affecting the growth of other plants species. Selectivity may be due to translocation, differential absorption, physical (morphological) or physiological differences between plant species. 2,4-D, mecoprop, dicamba control many broadleaf weeds but remain ineffective against turfgrasses.

- Non-selective herbicides: These herbicides are not specific in acting against certain plant species and control all plant material with which they come into contact. They are used to clear industrial sites, waste ground, railways and railway embankments. Paraquat, glufosinate, glyphosate are non-selective herbicides.

Timing of Application

- Preplant: Preplant herbicides are nonselective herbicides applied to soil before planting. Some preplant herbicides may be mechanically incorporated into the soil. The objective for incorporation is to prevent dissipation through photodecomposition and/or volatility. The herbicides kill weeds as they grow through the herbicide treated zone. Volatile herbicides have to be incorporated into the soil before planting the pasture. Agricultural crops grown in soil treated with a preplant herbicide include tomatoes, corn, soybeans and strawberries. Soil fumigants like metam-sodium and dazomet are in use as preplant herbicides.

- Preemergence: Preemergence herbicides are applied before the weed seedlings emerge through the soil surface. Herbicides do not prevent weeds from germinating but they kill weeds as they grow through the herbicide treated zone by affecting the cell division in the emerging seedling. Dithopyr and pendimethalin are preemergence herbicides. Weeds that have already emerged before application or activation are not affected by pre-herbicides as their primary growing point escapes the treatment.

- Postemergence: These herbicides are applied after weed seedlings have emerged through the soil surface. They can be foliar or root absorbed, selective or non-selective, contact or systemic. Application of these herbicides is avoided during rain because the problem of being washed off to the soil makes it ineffective. 2,4-D is a selective, systemic, foliar absorbed postemergence herbicide.

Method of Application

- Soil applied: Herbicides applied to the soil are usually taken up by the root or shoot of the emerging seedlings and are used as preplant or preemergence treatment. Several factors influence the effectiveness of soil-applied herbicides. Weeds absorb herbicides by both passive and active mechanism. Herbicide adsorption to soil colloids or organic matter often reduces its amount available for weed absorption. Positioning of herbicide in correct layer of soil is very important, which can be achieved mechanically and by rainfall. Herbicides on the soil surface are subjected to several processes that reduce their availability. Volatility and photolysis are two common processes that reduce the availability of herbicides. Many soil applied herbicides are absorbed through plant shoots while they are still underground leading to their death or injury. EPTC and trifluralin are soil applied herbicides.

- Foliar applied: These are applied to portion of the plant above the ground and are absorbed by exposed tissues. These are generally postemergence herbicides and can either be translocated (systemic) throughout the plant or remain at specific site (contact). External barriers of plants like cuticle, waxes, cell wall etc. affect herbicide absorption and action. Glyphosate, 2,4-D and dicamba are foliar applied herbicide.

Persistence

- Residual activity: A herbicide is described as having low residual activity if it is neutralized within a short time of application (within a few weeks or months) - typically this is due to rainfall, or by reactions in the soil. A herbicide described as having high residual activity will remains potent for a long term in the soil. For some compounds, the residual activity can leave the ground almost permanently barren.

Mechanism of Action

Herbicides are often classified according to their site of action, because as a general rule, herbicides within the same site of action class will produce similar symptoms on susceptible plants. Classification based on site of action of herbicide is comparatively better as herbicide resistance management can be handled more properly and effectively. Classification by mechanism of action (MOA) indicates the first enzyme, protein, or biochemical step affected in the plant following application.

List of Mechanisms Found in Modern Herbicides

- ACCase inhibitors compounds kill grasses. Acetyl coenzyme A carboxylase (ACCase) is part of the first step of lipid synthesis. Thus, ACCase inhibitors affect cell membrane production in the meristems of the grass plant. The ACCases of grasses are sensitive to these herbicides, whereas the ACCases of dicot plants are not.

- ALS inhibitors: the acetolactate synthase (ALS) enzyme (also known as acetohydroxyacid synthase, or AHAS) is the first step in the synthesis of the branched-chain amino acids (valine, leucine, and isoleucine). These herbicides slowly starve affected plants of these amino acids, which eventually leads to inhibition of DNA synthesis. They affect grasses and dicots alike. The ALS inhibitor family includes various sulfonylureas (such as Flazasulfuron and Metsulfuron-methyl), imidazolinones, triazolopyrimidines, pyrimidinyl oxybenzoates, and sulfonylamino carbonyl triazolinones. The ALS biological pathway exists only in plants and not animals, thus making the ALS-inhibitors among the safest herbicides.

- EPSPS inhibitors: The enolpyruvylshikimate 3-phosphate synthase enzyme EPSPS is used in the synthesis of the amino acids tryptophan, phenylalanine and tyrosine. They affect grasses and dicots alike. Glyphosate (Roundup) is a systemic EPSPS inhibitor inactivated by soil contact.

- Synthetic auxins inaugurated the era of organic herbicides. They were discovered in the 1940s after a long study of the plant growth regulator auxin. Synthetic auxins mimic this plant hormone. They have several points of action on the cell membrane, and are effective in the control of dicot plants. 2,4-D is a synthetic auxin herbicide.

- Photosystem II inhibitors reduce electron flow from water to NADPH2+ at the photochemical step in photosynthesis. They bind to the Qb site on the D1 protein, and prevent quinone from binding to this site. Therefore, this group of compounds causes electrons to accumulate on chlorophyll molecules. As a consequence, oxidation reactions in excess of those normally tolerated by the cell occur, and the plant dies. The triazine herbicides (including atrazine) and urea derivatives (diuron) are photosystem II inhibitors.

- Photosystem I inhibitors steal electrons from the normal pathway through FeS to Fdx to NADP leading to direct discharge of electrons on oxygen. As a result, reactive oxygen species are produced and oxidation reactions in excess of those normally tolerated by the cell occur, leading to plant death. Bipyridinium herbicides (such as diquat and paraquat) inhibit the Fe-S – Fdx step of that chain, while diphenyl ether herbicides (such as nitrofen, nitrofluorfen, and acifluorfen) inhibit the Fdx – NADP step.

- HPPD inhibitors inhibit 4-Hydroxyphenylpyruvate dioxygenase, which are involved in tyrosine breakdown. Tyrosine breakdown products are used by plants to make carotenoids, which protect chlorophyll in plants from being destroyed by sunlight. If this happens, the plants turn white due to complete loss of chlorophyll, and the plants die. Mesotrione and sulcotrione are herbicides in this class; a drug, nitisinone, was discovered in the course of developing this class of herbicides.

Herbicide Group (Labeling)

One of the most important methods for preventing, delaying, or managing resistance is to reduce the reliance on a single herbicide mode of action. To do this, farmers must know the mode of action for the herbicides they intend to use, but the relatively complex nature of plant biochemistry makes this difficult to determine. Attempts were made to simplify the understanding of herbicide mode of action by developing a classification system that grouped herbicides by mode of action. Eventually the Herbicide Resistance Action Committee (HRAC) and the Weed Science Society of America (WSSA) developed a classification system. The WSSA and HRAC systems differ in the group designation. Groups in the WSSA and the HRAC systems are designated by numbers and letters, respectively. The goal for adding the "Group" classification and mode of action to the herbicide product label is to provide a simple and practical approach to deliver the information to users. This information will make it easier to develop educational material that is consistent and effective. It should increase user's awareness of herbicide mode of action and provide more accurate recommendations for resistance management. Another goal is to make it easier for users to keep records on which herbicide mode of actions are being used on a particular field from year to year.

Chemical Family

Detailed investigations on chemical structure of the active ingredients of the registered herbicides showed that some moieties (moiety is a part of a molecule that may include either whole functional groups or parts of functional groups as substructures; a functional group has similar chemical properties whenever it occurs in different compounds) have the same mechanisms of action. According to Forouzesh *et al.* 2015, these moieties have been assigned to the names of chemical families and active ingredients are then classified within the chemical families accordingly. Knowing about herbicide chemical family grouping could serve as a short-term strategy for managing resistance to site of action.

Use and Application

Herbicides being sprayed from the spray arms of a tractor in North Dakota.

Most herbicides are applied as water-based sprays using ground equipment. Ground equipment varies in design, but large areas can be sprayed using self-propelled sprayers equipped with long booms, of 60 to 120 feet (18 to 37 m) with spray nozzles spaced every 20–30 inches (510–760 mm) apart. Towed, handheld, and even horse-drawn sprayers are also used. On large areas, herbicides may also at times be applied aerially using helicopters or airplanes, or through irrigation systems (known as chemigation).

A further method of herbicide application developed around 2010, involves ridding the soil of its active weed seed bank rather than just killing the weed. This can successfully treat annual plants but not perennials. Researchers at the Agricultural Research Service found that the application of herbicides to fields late in the weeds' growing season greatly reduces their seed production, and therefore fewer weeds will return the following season. Because most weeds are annuals, their seeds will only survive in soil for a year or two, so this method will be able to destroy such weeds after a few years of herbicide application.

Weed-wiping may also be used, where a wick wetted with herbicide is suspended from a boom and dragged or rolled across the tops of the taller weed plants. This allows treatment of taller grassland weeds by direct contact without affecting related but desirable shorter plants in the grassland sward beneath. The method has the benefit of avoiding spray drift. In Wales, a scheme offering free weed-wiper hire was launched in 2015 in an effort to reduce the levels of MCPA in water courses.

Misuse and Misapplication

Herbicide volatilisation or spray drift may result in herbicide affecting neighboring fields or plants, particularly in windy conditions. Sometimes, the wrong field or plants may be sprayed due to error.

Use Politically, Militarily, and in Conflict

Health and Environmental Effects

Herbicides have widely variable toxicity in addition to acute toxicity from occupational exposure levels.

Some herbicides cause a range of health effects ranging from skin rashes to death. The pathway of attack can arise from intentional or unintentional direct consumption, improper application resulting in the herbicide coming into direct contact with people or wildlife, inhalation of aerial sprays, or food consumption prior to the labeled preharvest interval. Under some conditions, certain herbicides can be transported via leaching or surface runoff to contaminate groundwater or distant surface water sources. Generally, the conditions that promote herbicide transport include intense storm events (particularly shortly after application) and soils with limited capacity to adsorb or retain the herbicides. Herbicide properties that increase likelihood of transport include persistence (resistance to degradation) and high water solubility.

Phenoxy herbicides are often contaminated with dioxins such as TCDD; research has suggested such contamination results in a small rise in cancer risk after occupational exposure to these herbicides. Triazine exposure has been implicated in a likely relationship to increased risk of breast cancer, although a causal relationship remains unclear.

Herbicide manufacturers have at times made false or misleading claims about the safety of their products. Chemical manufacturer Monsanto Company agreed to change its advertising after pressure from New York attorney general Dennis Vacco; Vacco complained about misleading claims that its spray-on glyphosate-based herbicides, including Roundup, were safer than table salt and "practically non-toxic" to mammals, birds, and fish (though proof that this was ever said is hard to find). Roundup is toxic and has resulted in death after being ingested in quantities ranging from 85 to 200 ml, although it has also been ingested in quantities as large as 500 ml with only mild or moderate symptoms. The manufacturer of Tordon 101 (Dow AgroSciences, owned by the Dow Chemical Company) has claimed Tordon 101 has no effects on animals and insects, in spite of evidence of strong carcinogenic activity of the active ingredient Picloram in studies on rats.

The risk of Parkinson's disease has been shown to increase with occupational exposure to herbicides and pesticides. The herbicide paraquat is suspected to be one such factor.

All commercially sold, organic and nonorganic herbicides must be extensively tested prior to approval for sale and labeling by the Environmental Protection Agency. However, because of the large number of herbicides in use, concern regarding health effects is significant. In addition to health effects caused by herbicides themselves, commercial herbicide mixtures often contain other chemicals, including inactive ingredients, which have negative impacts on human health.

Ecological Effects

Commercial herbicide use generally has negative impacts on bird populations, although the impacts are highly variable and often require field studies to predict accurately. Laboratory studies have at times overestimated negative impacts on birds due to toxicity, predicting serious problems that were not observed in the field. Most observed effects are due not to toxicity, but to habitat changes and the decreases in abundance of species on which birds rely for food or shelter. Herbicide use in silviculture, used to favor certain types of growth following clearcutting, can cause significant drops in bird populations. Even when herbicides which have low toxicity to birds are used, they decrease the abundance of many types of vegetation on which the birds rely. Herbicide use in agriculture in Britain has been linked to a decline in seed-eating bird species which rely on the weeds killed by the herbicides. Heavy use of herbicides in neotropical agricultural areas has been one of many factors implicated in limiting the usefulness of such agricultural land for wintering migratory birds.

Frog populations may be affected negatively by the use of herbicides as well. While some studies have shown that atrazine may be a teratogen, causing demasculinization in male frogs, the U.S. Environmental Protection Agency (EPA) and its independent Scientific Advisory Panel (SAP) examined all available studies on this topic and concluded that "atrazine does not adversely affect amphibian gonadal development based on a review of laboratory and field studies."

Scientific Uncertainty of Full Extent of Herbicide Effects

The health and environmental effects of many herbicides is unknown, and even the scientific community often disagrees on the risk. For example, a 1995 panel of 13 scientists reviewing studies on the carcinogenicity of 2,4-D had divided opinions on the likelihood 2,4-D causes cancer in humans. As of 1992, studies on phenoxy herbicides were too few to accurately assess the risk of many types of cancer from these herbicides, even though evidence was stronger that exposure to these herbicides is associated with increased risk of soft tissue sarcoma and non-Hodgkin lymphoma. Furthermore, there is some suggestion that herbicides can play a role in sex reversal of certain organisms that experience temperature-dependent sex determination, which could theoretically alter sex ratios.

Resistance

Weed resistance to herbicides has become a major concern in crop production worldwide. Resistance to herbicides is often attributed to lack of rotational programmes of herbicides and to continuous applications of herbicides with the same sites of action. Thus, a true understanding of the sites of action of herbicides is essential for strategic planning of herbicide-based weed control.

Plants have developed resistance to atrazine and to ALS-inhibitors, and more recently, to glyphosate herbicides. Marestail is one weed that has developed glyphosate resistance. Glyphosate-resistant weeds are present in the vast majority of soybean, cotton and corn farms in some U.S. states. Weeds that can resist multiple other herbicides are spreading. Few new herbicides are near commercialization, and none with a molecular mode of action for which there is no resistance. Because most herbicides could not kill all weeds, farmers rotated crops and herbicides to stop resistant weeds. During its initial years, glyphosate was not subject to resistance and allowed farmers to reduce the use of rotation.

A family of weeds that includes waterhemp (Amaranthus rudis) is the largest concern. A 2008-9 survey of 144 populations of waterhemp in 41 Missouri counties revealed glyphosate resistance in 69%. Weeds from some 500 sites throughout Iowa in 2011 and 2012 revealed glyphosate resistance in approximately 64% of waterhemp samples. The use of other killers to target "residual" weeds has become common, and may be sufficient to have stopped the spread of resistance From 2005 through 2010 researchers discovered 13 different weed species that had developed resistance to glyphosate. But since then only two more have been discovered. Weeds resistant to multiple herbicides with completely different biological action modes are on the rise. In Missouri, 43% of samples were resistant to two different herbicides; 6% resisted three; and 0.5% resisted four. In Iowa 89% of waterhemp samples resist two or more herbicides, 25% resist three, and 10% resist five.

For southern cotton, herbicide costs has climbed from between $50 and $75 per hectare a few years ago to about $370 per hectare in 2013. Resistance is contributing to a massive shift away from growing cotton; over the past few years, the area planted with cotton has declined by 70% in Arkansas and by 60% in Tennessee. For soybeans in Illinois, costs have risen from about $25 to $160 per hectare.

Dow, Bayer CropScience, Syngenta and Monsanto are all developing seed varieties resistant to herbicides other than glyphosate, which will make it easier for farmers to use alternative weed killers. Even though weeds have already evolved some resistance to those herbicides, Powles says the new seed-and-herbicide combos should work well if used with proper rotation.

Biochemistry of Resistance

Resistance to herbicides can be based on one of the following biochemical mechanisms:

- Target-site resistance: This is due to a reduced (or even lost) ability of the herbicide to bind to its target protein. The effect usually relates to an enzyme with a crucial function in a metabolic pathway, or to a component of an electron-transport system. Target-site resistance may also be caused by an overexpression of the target enzyme (via gene amplification or changes in a gene promoter).

- Non-target-site resistance: This is caused by mechanisms that reduce the amount of herbicidal active compound reaching the target site. One important mechanism is an enhanced metabolic detoxification of the herbicide in the weed, which leads to insufficient amounts of the active substance reaching the target site. A reduced uptake and translocation, or sequestration of the herbicide, may also result in an insufficient herbicide transport to the target site.

- Cross-resistance: In this case, a single resistance mechanism causes resistance to several herbicides. The term target-site cross-resistance is used when the herbicides bind to the same target site, whereas non-target-site cross-resistance is due to a single non-target-site mechanism (e.g., enhanced metabolic detoxification) that entails resistance across herbicides with different sites of action.

- Multiple resistance: In this situation, two or more resistance mechanisms are present within individual plants, or within a plant population.

Resistance Management

Worldwide experience has been that farmers tend to do little to prevent herbicide resistance developing, and only take action when it is a problem on their own farm or neighbor's. Careful observation is important so that any reduction in herbicide efficacy can be detected. This may indicate evolving resistance. It is vital that resistance is detected at an early stage as if it becomes an acute, whole-farm problem, options are more limited and greater expense is almost inevitable. Table 1 lists factors which enable the risk of resistance to be assessed. An essential pre-requisite for confirmation of resistance is a good diagnostic test. Ideally this should be rapid, accurate, cheap and accessible. Many diagnostic tests have been developed, including glasshouse pot assays, petri dish assays and chlorophyll fluorescence. A key component of such tests is that the response of the suspect population to a herbicide can be compared with that of known susceptible and resistant standards under controlled conditions. Most cases of herbicide resistance are a consequence of the repeated use of herbicides, often in association with crop monoculture and reduced cultivation practices. It is necessary, therefore, to modify these practices in order to prevent or delay the onset of resistance or to control existing resistant populations. A key objective should be the reduction in selection pressure. An integrated weed management (IWM) approach is required, in which as many tactics as possible are used to combat weeds. In this way, less reliance is placed on herbicides and so selection pressure should be reduced.

Optimising herbicide input to the economic threshold level should avoid the unnecessary use of herbicides and reduce selection pressure. Herbicides should be used to their greatest potential by ensuring that the timing, dose, application method, soil and climatic conditions are optimal for good activity. In the UK, partially resistant grass weeds such as *Alopecurus myosuroides* (blackgrass) and *Avena* spp. (wild oat) can often be controlled adequately when herbicides are applied at the 2-3 leaf stage, whereas later applications at the 2-3 tiller stage can fail badly. Patch spraying, or applying herbicide to only the badly infested areas of fields, is another means of reducing total herbicide use.

Table 1. Agronomic factors influencing the risk of herbicide resistance development

Factor	Low risk	High risk
Cropping system	Good rotation	Crop monoculture
Cultivation system	Annual ploughing	Continuous minimum tillage
Weed control	Cultural only	Herbicide only
Herbicide use	Many modes of action	Single modes of action
Control in previous years	Excellent	Poor
Weed infestation	Low	High
Resistance in vicinity	Unknown	Common

Approaches to Treating Resistant Weeds

Alternative Herbicides

When resistance is first suspected or confirmed, the efficacy of alternatives is likely to be the first consideration. The use of alternative herbicides which remain effective on resistant populations can be a successful strategy, at least in the short term. The effectiveness of alternative herbicides will be highly dependent on the extent of cross-resistance. If there is resistance to a single group of herbicides, then the use of herbicides from other groups may provide a simple and effective solution, at least in the short term. For example, many triazine-resistant weeds have been readily controlled by the use of alternative herbicides such as dicamba or glyphosate. If resistance extends to more than one herbicide group, then choices are more limited. It should not be assumed that resistance will automatically extend to all herbicides with the same mode of action, although it is wise to assume this until proved otherwise. In many weeds the degree of cross-resistance between the five groups of ALS inhibitors varies considerably. Much will depend on the resistance mechanisms present, and it should not be assumed that these will necessarily be the same in different populations of the same species. These differences are due, at least in part, to the existence of different mutations conferring target site resistance. Consequently, selection for different mutations may result in different patterns of cross-resistance. Enhanced metabolism can affect even closely related herbicides to differing degrees. For example, populations of *Alopecurus myosuroides* (blackgrass) with an enhanced metabolism mechanism show resistance to pendimethalin but not to trifluralin, despite both being dinitroanilines. This is due to differences in the vulnerability of these two herbicides to oxidative metabolism. Consequently, care is needed when trying to predict the efficacy of alternative herbicides.

Mixtures and Sequences

The use of two or more herbicides which have differing modes of action can reduce the selection for resistant genotypes. Ideally, each component in a mixture should:

- Be active at different target sites

- Have a high level of efficacy

- Be detoxified by different biochemical pathways

- Have similar persistence in the soil (if it is a residual herbicide)

- Exert negative cross-resistance

- Synergise the activity of the other component

No mixture is likely to have all these attributes, but the first two listed are the most important. There is a risk that mixtures will select for resistance to both components in the longer term. One practical advantage of sequences of two herbicides compared with mixtures is that a better appraisal of the efficacy of each herbicide component is possible, provided that sufficient time elapses between each application. A disadvantage with sequences is that two separate applications have to be made and it is possible that the later application will be less effective on weeds surviving the first application. If these are resistant, then the second herbicide in the sequence may increase selection for resistant individuals by killing the susceptible plants which were damaged but not killed by the first application, but allowing the larger, less affected, resistant plants to survive. This has been cited as one reason why ALS-resistant *Stellaria media* has evolved in Scotland recently (2000), despite the regular use of a sequence incorporating mecoprop, a herbicide with a different mode of action.

Herbicide Rotations

Rotation of herbicides from different chemical groups in successive years should reduce selection for resistance. This is a key element in most resistance prevention programmes. The value of this approach depends on the extent of cross-resistance, and whether multiple resistance occurs owing to the presence of several different resistance mechanisms. A practical problem can be the lack of awareness by farmers of the different groups of herbicides that exist. In Australia a scheme has been introduced in which identifying letters are included on the product label as a means of enabling farmers to distinguish products with different modes of action.

Farming Practices and Resistance: a Case Study

Herbicide resistance became a critical problem in Australian agriculture, after many Australian sheep farmers began to exclusively grow wheat in their pastures in the 1970s. Introduced varieties of ryegrass, while good for grazing sheep, compete intensely with wheat. Ryegrasses produce so many seeds that, if left unchecked, they can completely choke a field. Herbicides provided excellent control, while reducing soil disrupting because of less need to plough. Within little more than a decade, ryegrass and other weeds began to develop resistance. In response Australian farmers changed methods. By 1983, patches of ryegrass had become immune to Hoegrass, a family of herbicides that inhibit an enzyme called acetyl coenzyme A carboxylase.

Ryegrass populations were large, and had substantial genetic diversity, because farmers had planted many varieties. Ryegrass is cross-pollinated by wind, so genes shuffle frequently. To control its distribution farmers sprayed inexpensive Hoegrass, creating selection pressure. In addition, farmers sometimes diluted the herbicide in order to save money, which allowed some plants to survive application. When resistance appeared farmers turned to a group of herbicides that block acetolactate synthase. Once again, ryegrass in Australia evolved a kind of "cross-resistance" that allowed it to rapidly break down a variety of herbicides. Four classes of herbicides become ineffective within a few years. In 2013 only two herbicide classes, called Photosystem II and long-chain fatty acid inhibitors, were effective against ryegrass.

List of Common Herbicides

Synthetic Herbicides

- 2,4-D is a broadleaf herbicide in the phenoxy group used in turf and no-till field crop production. Now, it is mainly used in a blend with other herbicides to allow lower rates of herbicides to be used; it is the most widely used herbicide in the world, and third most commonly used in the United States. It is an example of synthetic auxin (plant hormone).

- Aminopyralid is a broadleaf herbicide in the pyridine group, used to control weeds on grassland, such as docks, thistles and nettles. It is notorious for its ability to persist in compost.

- Atrazine, a triazine herbicide, is used in corn and sorghum for control of broadleaf weeds and grasses. Still used because of its low cost and because it works well on a broad spectrum of weeds common in the US corn belt, atrazine is commonly used with other herbicides to reduce the overall rate of atrazine and to lower the potential for groundwater contamination; it is a photosystem II inhibitor.

- Clopyralid is a broadleaf herbicide in the pyridine group, used mainly in turf, rangeland, and for control of noxious thistles. Notorious for its ability to persist in compost, it is another example of synthetic auxin.

- Dicamba, a postemergent broadleaf herbicide with some soil activity, is used on turf and field corn. It is another example of a synthetic auxin.

- Glufosinate ammonium, a broad-spectrum contact herbicide, is used to control weeds after the crop emerges or for total vegetation control on land not used for cultivation.

- Fluazifop (Fuselade Forte), a post emergence, foliar absorbed, translocated grass-selective herbicide with little residual action. It is used on a very wide range of broad leaved crops for control of annual and perennial grasses.

- Fluroxypyr, a systemic, selective herbicide, is used for the control of broad-leaved weeds in small grain cereals, maize, pastures, rangeland and turf. It is a synthetic auxin. In cereal growing, fluroxypyr's key importance is control of cleavers, *Galium aparine*. Other key broadleaf weeds are also controlled.

- Glyphosate, a systemic nonselective herbicide, is used in no-till burndown and for weed control in crops genetically modified to resist its effects. It is an example of an EPSPs inhibitor.

- Imazapyr a nonselective herbicide, is used for the control of a broad range of weeds, including terrestrial annual and perennial grasses and broadleaf herbs, woody species, and riparian and emergent aquatic species.

- Imazapic, a selective herbicide for both the pre- and postemergent control of some annual and perennial grasses and some broadleaf weeds, kills plants by inhibiting the production of branched chain amino acids (valine, leucine, and isoleucine), which are necessary for protein synthesis and cell growth.

- Imazamox, an imidazolinone manufactured by BASF for postemergence application that is an acetolactate synthase (ALS) inhibitor. Sold under trade names Raptor, Beyond, and Clearcast.

- Linuron is a nonselective herbicide used in the control of grasses and broadleaf weeds. It works by inhibiting photosynthesis.

- MCPA (2-methyl-4-chlorophenoxyacetic acid) is a phenoxy herbicide selective for broadleaf plants and widely used in cereals and pasture.

- Metolachlor is a pre-emergent herbicide widely used for control of annual grasses in corn and sorghum; it has displaced some of the atrazine in these uses.

- Paraquat is a nonselective contact herbicide used for no-till burndown and in aerial destruction of marijuana and coca plantings. It is more acutely toxic to people than any other herbicide in widespread commercial use.

- Pendimethalin, a pre-emergent herbicide, is widely used to control annual grasses and some broad-leaf weeds in a wide range of crops, including corn, soybeans, wheat, cotton, many tree and vine crops, and many turfgrass species.

- Picloram, a pyridine herbicide, mainly is used to control unwanted trees in pastures and edges of fields. It is another synthetic auxin.

- Sodium chlorate *(disused/banned in some countries)*, a nonselective herbicide, is considered phytotoxic to all green plant parts. It can also kill through root absorption.

- Triclopyr, a systemic, foliar herbicide in the pyridine group, is used to control broadleaf weeds while leaving grasses and conifers unaffected.

- Several sulfonylureas, including Flazasulfuron and Metsulfuron-methyl, which act as ALS inhibitors and in some cases are taken up from the soil via the roots.

Organic Herbicides

Recently, the term "organic" has come to imply products used in organic farming. Under this definition, an organic herbicide is one that can be used in a farming enterprise that has been classified as organic. Commercially sold organic herbicides are expensive and may not be affordable for commercial farming. Depending on the application, they may be less effective than synthetic herbicides and are generally used along with cultural and mechanical weed control practices.

Homemade organic herbicides include:

- Corn gluten meal (CGM) is a natural pre-emergence weed control used in turfgrass, which reduces germination of many broadleaf and grass weeds.

- Vinegar is effective for 5–20% solutions of acetic acid, with higher concentrations most effective, but it mainly destroys surface growth, so respraying to treat regrowth is needed. Resistant plants generally succumb when weakened by respraying.

- Steam has been applied commercially, but is now considered uneconomical and inadequate. It controls surface growth but not underground growth and so re-spraying to treat regrowth of perennials is needed.

- Flame is considered more effective than steam, but suffers from the same difficulties.

- D-limonene (citrus oil) is a natural degreasing agent that strips the waxy skin or cuticle from weeds, causing dehydration and ultimately death.

- Saltwater or salt applied in appropriate strengths to the rootzone will kill most plants.

- Monocerin produced by certain fungi will kill certain weeds such as Johnson grass.

Of Historical Interest and Other

- 2,4,5-Trichlorophenoxyacetic acid (2,4,5-T) was a widely used broadleaf herbicide until being phased out starting in the late 1970s. While 2,4,5-T itself is of only moderate toxicity, the manufacturing process for 2,4,5-T contaminates this chemical with trace amounts of 2,3,7,8-tetrachlorodibenzo-p-dioxin (TCDD). TCDD is extremely toxic to humans. With proper temperature control during production of 2,4,5-T, TCDD levels can be held to about .005 ppm. Before the TCDD risk was well understood, early production facilities lacked proper temperature controls. Individual batches tested later were found to have as much as 60 ppm of TCDD. 2,4,5-T was withdrawn from use in the USA in 1983, at a time of heightened public sensitivity about chemical hazards in the environment. Public concern about dioxins was high, and production and use of other (non-herbicide) chemicals potentially containing TCDD contamination was also withdrawn. These included pentachlorophenol (a wood preservative) and PCBs (mainly used as stabilizing agents in transformer oil). Some feel that the 2,4,5-T withdrawal was not based on sound science. 2,4,5-T has since largely been replaced by dicamba and triclopyr.

- Agent Orange was a herbicide blend used by the British military during the Malayan Emergency and the U.S. military during the Vietnam War between January 1965 and April 1970 as a defoliant. It was a 50/50 mixture of the n-butyl esters of 2,4,5-T and 2,4-D. Because of TCDD contamination in the 2,4,5-T component, it has been blamed for serious illnesses in many people who were exposed to it. However, research on populations exposed to its dioxin contaminant have been inconsistent and inconclusive.

- Diesel, and other heavy oil derivatives, are known to be informally used at times, but are usually banned for this purpose.

Fertigation

Fertigation using white poly bag

Fertigation is the injection of fertilizers, soil amendments, and other water-soluble products into an irrigation system.

Fertigation is related to chemigation, the injection of chemicals into an irrigation system. The two terms are sometimes used interchangeably however chemigation is generally a more controlled and regulated process due to the nature of the chemicals used. Chemigation often involves pesticides, herbicides, and fungicides, some of which pose health threat to humans, animals, and the environment.

Uses

Fertigation is practiced extensively in commercial agriculture and horticulture. Fertigation is also increasingly being used for landscaping as dispenser units become more reliable and easier to use. Fertigation is used to add additional nutrients or to correct nutrient deficiencies detected in plant tissue analysis. It is usually practiced on high-value crops such as vegetables, turf, fruit trees, and ornamentals.

Commonly Used Nutrients

Most plant nutrients can be applied through irrigation systems. Nitrogen is the most commonly used plant nutrient. Naturally occurring nitrogen (N_2) is a diatomic molecule which makes up approximately 80% of the earth's atmosphere. Most plants cannot directly consume diatomic nitrogen, therefore nitrogen must be contained as a component of other chemical substances which plants can consume. Commonly, anhydrous ammonia, ammonium nitrate, and urea are used as bioavailable sources of nitrogen. Other nutrients needed by plants include phosphorus and potassium. Like nitrogen, plants require these substances to live but they must be contained in other chemical substances such as monoammonium phosphate or diammonium phosphate to serve as bioavailable nutrients. A common source of potassium is muriate of potash which is chemically potassium chloride. A soil fertility analysis is used to determine which of the more stable nutrients should be used.

Advantages

The benefits of fertigation methods over conventional or drop-fertilizing methods include:

- Increased nutrient absorption by plants.

- Reduction of fertilizer, chemicals, and water needed.

- Reduced leaching of chemicals into the water supply.

- Reduced water consumption due to the plant's increased root mass's ability to trap and hold water.

- Application of nutrients can be controlled at the precise time and rate necessary.

- Minimized risk of the roots contracting soil borne diseases through the contaminated soil.

- Reduction of soil erosion issues as the nutrients are pumped through the water drip system.

Disadvantages

- Concentration of the solution decreases as the fertilizer dissolves. This may lead to poor nutrient placement.

- The water supply for fertigation is to be kept separate from the domestic water supply to avoid contamination.

- Possible pressure loss in the main irrigation line.

- The process is dependent on the water supply's non-restriction by drought rationing.

Methods Used

- Drip irrigation-Less wasteful than sprinklers.

- Sprinkler systems-Increases leaf and fruit quality.

- Continuous application-Fertilizer is supplied at a constant rate.

- Three-stage application-Irrigation starts without fertilizers. Fertilizers are applied later in the process.

- Proportional application-Injection rate is proportional to water discharge rate.

- Quantitative application-Nutrient solution is applied in a calculated amount to each irrigation block.

- Other methods of application include the lateral move, the traveler gun, and solid set systems.

System Design

Fertigation assists distribution of fertilizers for farmers. The simplest type of fertigation system consists of a tank with a pump, distribution pipes, capillaries, and a dripper pen.

All systems should be placed on a raised or sealed platform, not in direct contact with the earth. Each system should also be fitted with chemical spill trays.

Because of the potential risk of contamination in the potable (drinking) water supply, a backflow prevention device is required for most fertigation systems. Backflow requirements may vary greatly. Therefore, it is very important to understand the proper level of backflow prevention required by law. In the United States, the minimum backflow protection is usually determined by state regulation. Each city or town may set the level of protection required.

Agricultural Lime

Agricultural lime, also called aglime, agricultural limestone, garden lime or liming, is a soil additive made from pulverized limestone or chalk. The primary active component is calcium carbonate. Additional chemicals vary depending on the mineral source and may include calcium oxide, magnesium oxide and magnesium carbonate. Unlike the types of lime called quicklime (calcium oxide) and slaked lime (calcium hydroxide), powdered limestone does not require lime burning in a lime kiln; it only requires milling.

The effects of agricultural lime on soil are:

- it increases the pH of acidic soil (the lower the pH the more acidic the soil); in other words, soil acidity is reduced and alkalinity increased

- it provides a source of calcium and magnesium for plants

- it permits improved water penetration for acidic soils

- it improves the uptake of major plant nutrients (nitrogen, phosphorus, and potassium) of plants growing on acid soils.

Lime may occur naturally in some soils but may require addition of sulfuric acid for its agricultural benefits to be realized. Gypsum is also used to supply calcium for plant nutrition. The concept of "corrected lime potential" to define the degree of base saturation in soils became the basis for procedures now used in soil testing laboratories to determine the "lime requirement" of soils.

Other forms of lime have common applications in agriculture and gardening, including dolomitic lime and hydrated lime. Dolomitic lime may be used as a soil input to provide similar effects as agricultural lime, while supplying magnesium in addition to calcium. In livestock farming, hydrated lime can be used as a disinfectant measure, producing a dry and alkaline environment in which bacteria do not readily multiply. In horticultural farming it can be used as an insect repellent, without causing harm to the pest or plant.

Spinner-style lime spreaders are generally used to spread agricultural lime on fields.

Agricultural lime is injected into coal burners at power plants to reduce the pollutants such as NO_2 and SO_2 from the emissions.

Determining the Need for Agricultural Lime

The primary reason to apply agricultural lime is to correct the high levels of acidity in the soil. Acid soils reduce plant growth by inhibiting the intake of major plant nutrients (nitrogen, phosphorus and potassium). Some plants, particularly legumes, will not grow in highly acidic soils. This is vital to maximise crop yield, animal grazing and good quality silage/hay.

Soils become acidic in a number of ways. Locations that have high rainfall levels become acidic through leaching. Land used for crop and livestock purposes lose minerals over time by crop removal and become acidic. For example, when a 600 pound calf is removed from a pasture, 100 pounds of bone is also removed, which is 60% calcium compounds. The application of modern chemical fertilizers is a major contributor to soil acid by the process in which the plant nutrients react in the soil.

Aglime, which is high in calcium, can also be beneficial to soils where the land is used for breeding and raising foraging animals. Bone growth is key to a young animal's development and bones are composed primarily of calcium and phosphorus. Young mammals get their needed calcium through milk, which has calcium as one of its major components. Dairymen frequently apply aglime because it increases milk production.

The best way to determine if a soil is acid or deficient in calcium or magnesium is with a soil test which can be provided by a university with an agricultural education department for under $30.00, if you live in the United States. Farmers typically become interested in soil testing when they notice a decrease in crop response to applied fertilizer.

Quality

The quality of agricultural limestone is determined by the chemical makeup of the limestone and how finely the stone is ground. To aid the farmer in determining the relative value of competing agricultural liming materials, the agricultural extension services of several universities use two rating systems. Calcium Carbonate Equivalent (CCE) and the Effective Calcium Carbonate Equivalent (ECCE) give a numeric value to the effectiveness of different liming materials.

The CCE compares the chemistry of a particular quarry's stone with the neutralizing power of pure calcium carbonate. Because each molecule of magnesium carbonate is lighter than calcium carbonate, limestones containing magnesium carbonate (dolomite) can have a CCE greater than 100 percent.

Because the acids in soil are relatively weak, agricultural limestones must be ground to a small particle size to be effective. The extension service of different states rate the effectiveness of stone size particles slightly differently. They all agree, however, that the smaller the particle size the more effective the stone is at reacting in the soil. Measuring the size of particles is based on the size of a mesh that the limestone would pass through. The mesh size is the number of wires per inch. Stone retained on an 8 mesh will be about the size of BB pellets. Material passing a 60 mesh screen will have the appearance of face powder. Particles larger than 8 mesh are of little or no value, particles between 8 mesh and 60 mesh are somewhat effective and particles smaller than 60 mesh are 100 percent effective.

By combining the chemistry of a particular product (CCE) and its particle size the Effective Calcium Carbonate Equivalent (ECCE) is determined. The ECCE is percentage comparison of a particular agricultural limestone with pure calcium carbonate with all particles smaller than 60 mesh. Typically the aglime materials in commercial use will have ECCE ranging from 45 percent to 110 percent.

Brazil's Case

Brazil's vast inland cerrado region was regarded as unfit for farming before the 1960s because the soil was too acidic and poor in nutrients, according to Nobel Peace Prize winner Norman Borlaug, an American plant scientist referred to as the father of the Green Revolution. However, from the 1960s, vast quantities of lime (pulverised chalk or limestone) were poured on the soil to reduce acidity. The effort went on and in the late 1990s between 14 million and 16 million tonnes of lime were being spread on Brazilian fields each year. The quantity rose to 25 million tonnes in 2003 and 2004, equalling around five tonnes of lime per hectare. As a result, Brazil has become the world's second biggest soybean exporter and, thanks to the boom in animal feed production, Brazil is now the biggest exporter of beef and poultry in the world.

Pest Control

A crop duster applies low-insecticide bait that is targeted against western corn rootworms.

Pest control refers to the regulation or management of a species defined as a pest, and can be perceived to be detrimental to a person's health, the ecology or the economy. A practitioner of pest control is called an exterminator.

History

Pest control is at least as old as agriculture, as there has always been a need to keep crops free from pests. In order to maximize food production, it is advantageous to protect crops from competing species of plants, as well as from herbivores competing with humans.

The conventional approach was probably the first to be employed, since it is comparatively easy to destroy weeds by burning them or plowing them under, and to kill larger competing herbivores, such as crows and other birds eating seeds. Techniques such as crop rotation, companion planting (also known as intercropping or mixed cropping), and the selective breeding of pest-resistant cultivars have a long history.

In the UK, following concern about animal welfare, humane pest control and deterrence is gaining ground through the use of animal psychology rather than destruction. For instance, with the urban red fox which territorial behaviour is used against the animal, usually in conjunction with non-injurious chemical repellents. In rural areas of Britain, the use of firearms for pest control is quite common. Airguns are particularly popular for control of small pests such as rats, rabbits and grey squirrels, because of their lower power they can be used in more restrictive spaces such as gardens, where using a firearm would be unsafe.

Chemical pesticides date back 4,500 years, when the Sumerians used sulfur compounds as insecticides. The Rig Veda, which is about 4,000 years old, also mentions the use of poisonous plants for pest control. It was only with the industrialization and mechanization of agriculture in the 18th and 19th century, and the introduction of the insecticides pyrethrum and derris that chemical pest control became widespread. In the 20th century, the discovery of several synthetic insecticides, such as DDT, and herbicides boosted this development. Chemical pest control is still the predominant type of pest control today, although its long-term effects led to a renewed interest in traditional and biological pest control towards the end of the 20th century.

Causes

Many pests have only become a problem as a result of the direct actions by humans. Modifying these actions can often substantially reduce the pest problem. In the United States, raccoons caused a nuisance by tearing open refuse sacks. Many householders introduced bins with locking lids, which deterred the raccoons from visiting. House flies tend to accumulate wherever there is human activity and live in close association with people all over the world especially where food or food waste is exposed. Similarly, seagulls have become pests at many seaside resorts. Tourists would often feed the birds with scraps of fish and chips, and before long, the birds would rely on this food source and act aggressively towards humans.

Living organisms evolve and increase their resistance to biological, chemical, physical or any other form of control. Unless the target population is completely exterminated or is rendered incapable of reproduction, the surviving population will inevitably acquire a tolerance of whatever pressures are brought to bear - this results in an evolutionary arms race.

Sign in Ilfracombe, England designed to help control seagull presence

Types of Pest Control

Use of Pest-destroying Animals

Perhaps as far ago as 3000BC in Egypt, cats were being used to control pests of grain stores such as rodents. In 1939/40 a survey discovered that cats could keep a farm's population of rats down to a low level, but could not eliminate them completely. However, if the rats were cleared by trapping or poisoning, farm cats could stop them returning - at least from an area of 50 yards around a barn.

Ferrets were domesticated at least by 500 AD in Europe, being used as mousers. Mongooses have been introduced into homes to control rodents and snakes, probably at first by the ancient Egyptians.

Biological Pest Control

Biological pest control is the control of one through the control and management of natural predators and parasites. For example: mosquitoes are often controlled by putting *Bt Bacillus thuringiensis* ssp. *israelensis*, a bacterium that infects and kills mosquito larvae, in local water sources. The treatment has no known negative consequences on the remaining ecology and is safe for humans to drink. The point of biological pest control, or any natural pest control, is to eliminate a pest with minimal harm to the ecological balance of the environment in its present form.

Mechanical Pest Control

Mechanical pest control is the use of hands-on techniques as well as simple equipment and devices, that provides a protective barrier between plants and insects. For example: weeds can be controlled by being physically removed from the ground. This is referred to as tillage and is one of the oldest methods of weed control.

Physical Pest Control

Dog control van, Rekong Peo, Himachal Pradesh, India

Physical pest control is a method of getting rid of insects and small rodents by removing, attacking, setting up barriers that will prevent further destruction of one's plants, or forcing insect infestations to become visual.

Elimination of Breeding Grounds

Proper waste management and drainage of still water, eliminates the breeding ground of many pests.

Garbage provides food and shelter for many unwanted organisms, as well as an area where still water might collect and be used as a breeding ground by mosquitoes. Communities that have proper garbage collection and disposal, have far less of a problem with rats, cockroaches, mosquitoes, flies and other pests than those that don't.

Open air sewers are ample breeding ground for various pests as well. By building and maintaining a proper sewer system, this problem is eliminated.

Certain spectrums of LED light can "disrupt insects' breeding".

Poisoned Bait

Poisoned bait is a common method for controlling rat populations, however is not as effective when there are other food sources around, such as garbage. Poisoned meats have been used for centuries for killing off wolves, birds that were seen to threaten crops, and against other creatures. This can be a problem, since a carcass which has been poisoned will kill not only the targeted animal, but also every other animal which feeds on the carcass. Humans have also been killed by coming in contact with poisoned meat, or by eating an animal which had fed on a poisoned carcass. This tool is also used to manage several caterpillars e.g. Spodoptera litura, fruit flies, snails and slugs, crabs etc.

Field Burning

Traditionally, after a sugar cane harvest, the fields are all burned, to kill off any rodents, insects or eggs that might be in the fields.

Hunting

Historically, in some European countries, when stray dogs and cats became too numerous, local populations gathered together to round up all animals that did not appear to have an owner and kill them. In some nations, teams of rat-catchers work at chasing rats from the field, and killing them with dogs and simple hand tools. Some communities have in the past employed a bounty system, where a town clerk will pay a set fee for every rat head brought in as proof of a rat killing.

Traps

A variety of mouse traps and rat traps are available for mice and rats, including snap traps, glue traps and live catch traps.

Pesticides

Rodent bait station, Chennai, India

Spraying pesticides by planes, trucks or by hand is a common method of pest control. Crop dusters commonly fly over farmland and spray pesticides to kill off pests that would threaten the crops. However, some pesticides may cause cancer and other health problems, as well as harming wildlife.

Space Fumigation

A project that involves a structure be covered or sealed airtight followed by the introduction of a penetrating, deadly gas at a killing concentration a long period of time (24-72hrs.). Although expensive, space fumigation targets all life stages of pests.

Space Treatment

Residential & commercial building pest control service vehicle, Ypsilanti Township, Michigan

A long term project involving fogging or misting type applicators. Liquid insecticide is dispersed in the atmosphere within a structure. Treatments do not require the evacuation or airtight sealing of a building, allowing most work within the building to continue but at the cost of the penetrating effects. Contact insecticides are generally used, minimizing the long lasting residual effects. On August 10, 1973, the Federal Register printed the definition of Space treatment as defined by the U.S. Environmental Protection Agency (EPA):

> the dispersal of insecticides into the air by foggers, misters, aerosol devices or vapor dispensers for control of flying insects and exposed crawling insects

Sterilization

Laboratory studies conducted with U-5897 (3-chloro-1,2-propanediol) were attempted in the early 1970s although these proved unsuccessful. Research into sterilization bait is ongoing.

In 2013, New York City tested sterilization traps in a $1.1 million study. The result was a 43% reduction in rat populations. The Chicago Transit Authority plans to test sterilization control in spring 2015. The sterilization method doesn't poison the rats or humans.

Destruction of Infected Plants

Forest services sometimes destroy all the trees in an area where some are infected with insects, if seen as necessary to prevent the insect species from spreading. Farms infested with certain insects, have been burned entirely, to prevent the pest from spreading elsewhere.

Natural Rodent Control

Several wildlife rehabilitation organizations encourage natural form of rodent control through exclusion and predator support and preventing secondary poisoning altogether.

Example of House mouse infestation

The United States Environmental Protection Agency agrees, noting in its Proposed Risk Mitigation Decision for Nine Rodenticides that "without habitat modification to make areas less attractive to commensal rodents, even eradication will not prevent new populations from recolonizing the habitat."

Repellents

- Balsam fir oil from the tree *Abies balsamea* is an EPA approved non-toxic rodent repellent.

- *Acacia polyacantha* subsp. *campylacantha* root emits chemical compounds that repel animals including crocodiles, snakes and rats.

Integrated Pest Management

An IPM boll weevil trap in a cotton field (Manning, South Carolina).

Integrated pest management (IPM), also known as integrated pest control (IPC) is a broad-based approach that integrates practices for economic control of pests. IPM aims to suppress pest populations below the economic injury level (EIL). The UN's Food and Agriculture Organisation defines IPM as "the careful consideration of all available pest control techniques and subsequent integration of appropriate measures that discourage the development of pest populations and keep pesticides and other interventions to levels that are economically justified and reduce or minimize risks to human health and the environment. IPM emphasizes the growth of a healthy crop with the least possible disruption to agro-ecosystems and encourages natural pest control mechanisms." Entomologists and ecologists have urged the adoption of IPM pest control since the 1970s. IPM allows for safer pest control. This includes managing insects, plant pathogens and weeds.

Globalization and increased mobility often allow increasing numbers of invasive species to cross national borders. IPM poses the least risks while maximizing benefits and reducing costs.

For their leadership in developing and spreading IPM worldwide, Perry Adkisson and Ray F. Smith received the 1997 World Food Prize.

History

Shortly after World War II, when synthetic insecticides became widely available, entomologists in California developed the concept of "supervised insect control". Around the same time, entomologists in the US Cotton Belt were advocating a similar approach. Under this scheme, insect control was "supervised" by qualified entomologists and insecticide applications were based on conclusions reached from periodic monitoring of pest and natural-enemy populations. This was viewed as an alternative to calendar-based programs. Supervised control was based on knowledge of the ecology and analysis of projected trends in pest and natural-enemy populations.

Supervised control formed much of the conceptual basis for the "integrated control" that University of California entomologists articulated in the 1950s. Integrated control sought to identify the best mix of chemical and biological controls for a given insect pest. Chemical insecticides were to be used in the manner least disruptive to biological control. The term "integrated" was thus synonymous with "compatible." Chemical controls were to be applied only after regular monitoring indicated that a pest population had reached a level (the economic threshold) that required treatment to prevent the population from reaching a level (the economic injury level) at which economic losses would exceed the cost of the control measures.

IPM extended the concept of integrated control to all classes of pests and was expanded to include all tactics. Controls such as pesticides were to be applied as in integrated control, but these now had to be compatible with tactics for all classes of pests. Other tactics, such as host-plant resistance and cultural manipulations, became part of the IPM framework. IPM combined entomologists, plant pathologists, nematologists and weed scientists.

In the United States, IPM was formulated into national policy in February 1972 when President Richard Nixon directed federal agencies to take steps to advance the application of IPM in all relevant sectors. In 1979, President Jimmy Carter established an interagency IPM Coordinating Committee to ensure development and implementation of IPM practices.

Applications

IPM is used in agriculture, horticulture, human habitations, preventive conservation and general pest control, including structural pest management, turf pest management and ornamental pest management.

Principles

An American IPM system is designed around six basic components:

- Acceptable pest levels—The emphasis is on *control*, not *eradication*. IPM holds that wiping out an entire pest population is often impossible, and the attempt can be expensive and unsafe. IPM programmes first work to establish acceptable pest levels, called action thresholds, and apply controls if those thresholds are crossed. These thresholds are pest and site specific, meaning that it may be acceptable at one site to have a weed such as white clover, but not at another site. Allowing a pest population to survive at a reasonable threshold reduces selection pressure. This lowers the rate at which a pest develops resistance to a control, because if almost all pests are killed then those that have resistance will provide the genetic basis of the future population. Retaining a significant number of unresistant specimens dilutes the prevalence of any resistant genes that appear. Similarly, the repeated use of a single class of controls will create pest populations that are more resistant to that class, whereas alternating among classes helps prevent this.

- Preventive cultural practices—Selecting varieties best for local growing conditions and maintaining healthy crops is the first line of defense. Plant quarantine and 'cultural techniques' such as crop sanitation are next, e.g., removal of diseased plants, and cleaning pruning shears to prevent spread of infections. Beneficial fungi and bacteria are added to the potting media of horticultural crops vulnerable to root diseases, greatly reducing the need for fungicides.

- Monitoring—Regular observation is critically important. Observation is broken into inspection and identification. Visual inspection, insect and spore traps, and other methods are used to monitor pest levels. Record-keeping is essential, as is a thorough knowledge target pest behavior and reproductive cycles. Since insects are cold-blooded, their physical development is dependent on area temperatures. Many insects have had their development cycles modeled in terms of degree-days. The degree days of an environment determines the optimal time for a specific insect outbreak. Plant pathogens follow similar patterns of response to weather and season.

- Mechanical controls—Should a pest reach an unacceptable level, mechanical methods are the first options. They include simple hand-picking, barriers, traps, vacuuming and tillage to disrupt breeding.

- Biological controls—Natural biological processes and materials can provide control, with acceptable environmental impact, and often at lower cost. The main approach is to promote beneficial insects that eat or parasitize target pests. Biological insecticides, derived from naturally occurring microorganisms (e.g.—Bt, entomopathogenic fungi and entomopathogenic nematodes), also fall in this category. Further 'biology-based' or 'ecological' techniques are under evaluation.

- Responsible use—Synthetic pesticides are used as required and often only at specific times in a pest's life cycle. Many newer pesticides are derived from plants or naturally occurring substances (e.g.—nicotine, pyrethrum and insect juvenile hormone analogues), but the toxophore or active component may be altered to provide increased biological activity or stability. Applications of pesticides must reach their intended targets. Matching the application technique to the crop, the pest, and the pesticide is critical. The use of low-volume spray equipment reduces overall pesticide use and labor cost.

An IPM regime can be simple or sophisticated. Historically, the main focus of IPM programmes was on agricultural insect pests. Although originally developed for agricultural pest management, IPM programmes are now developed to encompass diseases, weeds and other pests that interfere with management objectives for sites such as residential and commercial structures, lawn and turf areas, and home and community gardens.

Process

IPM is the selection and use of pest control actions that will ensure favourable economic, ecological and social consequences and is applicable to most agricultural, public health and amenity pest management situations. The IPM process starts with monitoring, which includes inspection and identification, followed by the establishment of economic injury levels. The economic injury levels set the economic threshold level. That is the point when pest damage (and the benefits of treating the pest) exceed the cost of treatment. This can also be an action threshold level for determining an unacceptable level that is not tied to economic injury. Action thresholds are more common in structural pest management and economic injury levels in classic agricultural pest management. An example of an action threshold is one fly in a hospital operating room is not acceptable, but one fly in a pet kennel would be acceptable. Once a threshold has been crossed by the pest population action steps need to be taken to reduce and control the pest. Integrated pest management employ a variety of actions including cultural controls, including physical barriers, biological controls, including adding and conserving natural predators and enemies to the pest, and finally chemical controls or pesticides. Reliance on knowledge, experience, observation and integration of multiple techniques makes IPM appropriate for organic farming (excluding synthetic pesticides). These may or may not include materials listed on the Organic Materials Review Institute (OMRI) Although the pesticides and particularly insecticides used in organic farming and organic gardening are generally safer than synthetic pesticides, they are not always more safe or environmentally friendly than synthetic pesticides and can cause harm. For conventional farms IPM can reduce human and environmental exposure to hazardous chemicals, and potentially lower overall costs.

Risk assessment usually includes four issues: 1) characterization of biological control agents, 2) health risks, 3) environmental risks and 4) efficacy.

Mistaken identification of a pest may result in ineffective actions. E.g., plant damage due to over-watering could be mistaken for fungal infection, since many fungal and viral infections arise under moist conditions.

Monitoring begins immediately, before the pest's activity becomes significant. Monitoring of agricultural pests includes tracking soil/planting media fertility and water quality. Overall plant health and resistance to pests is greatly influenced by pH, alkalinity, of dissolved mineral and Oxygen Reduction Potential. Many diseases are waterborne, spread directly by irrigation water and indirectly by splashing.

Once the pest is known, knowledge of its lifecycle provides the optimal intervention points. For example, weeds reproducing from last year's seed can be prevented with mulches and pre-emergent herbicide.

Pest-tolerant crops such as soybeans may not warrant interventions unless the pests are numerous or rapidly increasing. Intervention is warranted if the expected cost of damage by the pest is more than the cost of control. Health hazards may require intervention that is not warranted by economic considerations.

Specific sites may also have varying requirements. E.g., white clover may be acceptable on the sides of a tee box on a golf course, but unacceptable in the fairway where it could confuse the field of play.

Possible interventions include mechanical/physical, cultural, biological and chemical. Mechanical/physical controls include picking pests off plants, or using netting or other material to exclude pests such as birds from grapes or rodents from structures. Cultural controls include keeping an area free of conducive conditions by removing waste or diseased plants, flooding, sanding, and the use of disease-resistant crop varieties.

Biological controls are numerous. They include: conservation of natural predators or augmentation of natural predators, Sterile insect technique (SIT).

Augmentation, inoculative release and inundative release are different methods of biological control that affect the target pest in different ways. Augmentative control includes the periodic introduction of predators. With inundative release, predators are collected, mass-reared and periodically released in large numbers into the pest area. This is used for an immediate reduction in host populations, generally for annual crops, but is not suitable for long run use. With inoculative release a limited number of beneficial organisms are introduced at the start of the growing season. This strategy offers long term control as the organism's progeny affect pest populations throughout the season and is common in orchards. With seasonal inoculative release the beneficials are collected, mass-reared and released seasonally to maintain the beneficial population. This is commonly used in greenhouses. In America and other western countries, inundative releases are predominant, while Asia and the eastern Europe more commonly use inoculation and occasional introductions.

The Sterile insect technique (SIT) is an Area-Wide IPM program that introduces sterile male pests into the pest population to trick females into (unsuccessful) breeding encounters, providing a form of birth control and reducing reproduction rates. The biological controls mentioned above only appropriate in extreme cases, because in the introduction of new species, or supplementation of naturally occurring species can have detrimental ecosystem effects. Biological controls can be used to stop invasive species or pests, but they can become an introduction path for new pests.

Chemical controls include horticultural oils or the application of insecticides and herbicides. A Green Pest Management IPM program uses pesticides derived from plants, such as botanicals, or other naturally occurring materials.

Pesticides can be classified by their modes of action. Rotating among materials with different modes of action minimizes pest resistance.

Evaluation is the process of assessing whether the intervention was effective, whether it produced unacceptable side effects, whether to continue, revise or abandon the program.

Southeast Asia

The Green Revolution of the 1960s and '70s introduced sturdier plants that could support the heavier grain loads resulting from intensive fertilizer use. Pesticide imports by 11 Southeast Asian countries grew nearly sevenfold in value between 1990 and 2010, according to FAO statistics, with disastrous results. Rice farmers become accustomed to spraying soon after planting, triggered by signs of the leaf folder moth, which appears early in the growing season. It causes only superficial damage and doesn't reduce yields. In 1986, Indonesia banned 57 pesticides and completely stopped subsidizing their use. Progress was reversed in the 2000s, when growing production capacity, particularly in China, reduced prices. Rice production in Asia more than doubled. But it left farmers believing more is better—whether it's seed, fertilizer, or pesticides.

The brown planthopper (Nilaparvata lugens), the farmers' main target, has become increasingly resistant. Since 2008, outbreaks have devastated rice harvests throughout Asia, but not in the Mekong Delta. Reduced spraying allowed natural predators to neutralize planthoppers in Vietnam. In 2010 and 2011, massive planthopper outbreaks hit 400,000 hectares of Thai rice fields, causing losses of about $64 million. The Thai government is now pushing the "no spray in the first 40 days" approach.

By contrast early spraying kills frogs, spiders, wasps and dragonflies that prey on the later-arriving and dangerous planthopper and produced resistant strains. Planthoppers now require pesticide doses 500 times greater than originally. Overuse indiscriminately kills beneficial insects and decimates bird and amphibian populations. Pesticides are suspected of harming human health and became a common means for rural Asians to commit suicide.

In 2001, scientists challenged 950 Vietnamese farmers to try IPM. In one plot, each farmer grew rice using their usual amounts of seed and fertilizer, applying pesticide as they chose. In a nearby plot, less seed and fertilizer were used and no pesticides were applied for 40 days after planting. Yields from the experimental plots was as good or better and costs were lower, generating 8% to 10% more net income. The experiment led to the "three reductions, three gains" campaign, claiming that cutting the use of seed, fertilizer and pesticide would boost yield, quality and income. Posters, leaflets, TV commercials and a 2004 radio soap opera that featured a rice farmer who gradually accepted the changes. It didn't hurt that a 2006 planthopper outbreak hit farmers using insecticides harder than those who didn't. Mekong Delta farmers cut insecticide spraying from five times per crop cycle to zero to one.

The Plant Protection Center and the International Rice Research Institute (IRRI) have been encouraging farmers to grow flowers, okra and beans on rice paddy banks, instead of stripping vegetation, as was typical. The plants attract bees and a tiny wasp that eats planthopper eggs, while the vegetables diversify farm incomes.

Agriculture companies offer bundles of pesticides with seeds and fertilizer, with incentives for volume purchases. A proposed law in Vietnam requires licensing pesticide dealers and government approval of advertisements to prevent exaggerated claims. Insecticides that target other pests, such as Scirpophaga incertulas (stem borer), the larvae of moth species that feed on rice plants allegedly yield gains of 21% with proper use.

Biopesticide

Biopesticides, a contraction of 'biological pesticides', include several types of pest management intervention: through predatory, parasitic, or chemical relationships. The term has been associated historically with biological control - and by implication - the manipulation of living organisms. Regulatory positions can be influenced by public perceptions, thus:

- in the EU, biopesticides have been defined as "a form of pesticide based on micro-organisms or natural products".

- the US EPA states that they "include naturally occurring substances that control pests (biochemical pesticides), microorganisms that control pests (microbial pesticides), and pesticidal substances produced by plants containing added genetic material (plant-incorporated protectants) or PIPs".

They are obtained from organisms including plants, bacteria and other microbes, fungi, nematodes, *etc.* They are often important components of integrated pest management (IPM) programmes, and have received much practical attention as substitutes to synthetic chemical plant protection products (PPPs).

Types

Biopesticides can be classified into these classes:

- Microbial pesticides which consist of bacteria, entomopathogenic fungi or viruses (and sometimes includes the metabolites that bacteria or fungi produce). Entomopathogenic nematodes are also often classed as microbial pesticides, even though they are multi-cellular.

- Bio-derived chemicals. Four groups are in commercial use: pyrethrum, rotenone, neem oil, and various essential oils are naturally occurring substances that control (or monitor in the case of pheromones) pests and microbial diseases.

- Plant-incorporated protectants (PIPs) have genetic material from other species incorporated into their genetic material (*i.e.* GM crops). Their use is controversial, especially in many European countries.

- RNAi pesticides, some of which are topical and some of which are absorbed by the crop.

Biopesticides have usually no known function in photosynthesis, growth or other basic aspects of plant physiology. Instead, they are active against biological pests. Many chemical compounds have been identified that are produced by plants to protect them from pests. These materials are biodegradable and renewable alternatives, which can be economical for practical use. Organic farming systems embraces this approach to pest control.

RNAi

RNA interference is under study for possible use as a spray-on insecticide by multiple companies, including Monsanto, Syngenta, and Bayer. Such sprays do not modify the genome of the target plant. The RNA could be modified to maintain its effectiveness as target species evolve tolerance to the original. RNA is a relatively fragile molecule that generally degrades within days or weeks of application. Monsanto estimated costs to be on the order of $5/acre.

RNAi has been used to target weeds that tolerate Monsanto's Roundup herbicide. RNAi mixed with a silicone surfactant that let the RNA molecules enter air-exchange holes in the plant's surface that disrupted the gene for tolerance, affecting it long enough to let the herbicide work. This strategy would allow the continued use of glyphosate-based herbicides, but would not per se assist a herbicide rotation strategy that relied on alternating Roundup with others.

They can be made with enough precision to kill some insect species, while not harming others. Monsanto is also developing an RNA spray to kill potato beetles One challenge is to make it linger on the plant for a week, even if it's raining. The Potato beetle has become resistant to more than 60 conventional insecticides.

Monsanto lobbied the U.S. EPA to exempt RNAi pesticide products from any specific regulations (beyond those that apply to all pesticides) and be exempted from rodent toxicity, allergenicity and residual environmental testing. In 2014 an EPA advisory group found little evidence of a risk to people from eating RNA.

However, in 2012, the Australian Safe Food Foundation alleged that the RNA trigger designed to change wheat's starch content might interfere with the gene for a human liver enzyme. Supporters countered that RNA does not appear to make it past human saliva or stomach acids. The US National Honey Bee Advisory Board told EPA that using RNAi would put natural systems at "the epitome of risk". The beekeepers cautioned that pollinators could be hurt by unintended effects and that the genomes of many insects are still unknown. Other unassessed risks include ecological (given the need for sustained presence for herbicide and other applications) and the possible for RNA drift across species boundaries.

Monsanto has invested in multiple companies for their RNA expertise, including Beeologics (for RNA that kills a parasitic mite that infests hives and for manufacturing technology) and Preceres (nanoparticle lipidoid coatings) and licensed technology from Alnylam and Tekmira. In 2012 Syngenta acquired Devgen, a European RNA partner. Startup Forrest Innovations is investigating RNAi as a solution to citrus greening disease that in 2014 caused 22 percent of oranges in Florida to fall off the trees.

Examples

Bacillus thuringiensis, a bacterial disease of Lepidoptera, Coleoptera and Diptera, is a well-known insecticide example. The toxin from *B. thuringiensis* (Bt toxin) has been incorporated directly into plants through the use of genetic engineering. The use of Bt Toxin is particularly controversial. Its manufacturers claim it has little effect on other organisms, and is more environmentally friendly than synthetic pesticides. However, at least one scientific study has suggested that it may lead to slight histopathological changes on the liver and kidneys of mammals with Bt toxin in their diet.

Other microbial control agents include products based on:

- entomopathogenic fungi (*e.g.Beauveria bassiana*, Paecilomyces fumosoroseus, *Lecanicillium* spp., *Metarhizium* spp.),

- plant disease control agents: include *Trichoderma* spp. and *Ampelomyces quisqualis* (a hyper-parasite of grape powdery mildew); *Bacillus subtilis* is also used to control plant pathogens.

- beneficial nematodes attacking insect (*e.g. Steinernema feltiae*) or slug (*e.g. Phasmarhabditis hermaphrodita*) pests

- entomopathogenic viruses (*e.g.. Cydia pomonella* granulovirus).

- weeds and rodents have also been controlled with microbial agents.

Various naturally occurring materials, including fungal and plant extracts, have been described as biopesticides. Products in this category include:

- Insect pheromones and other semiochemicals

- Fermentation products such as Spinosad (a macro-cyclic lactone)

- Chitosan: a plant in the presence of this product will naturally induce systemic resistance (ISR) to allow the plant to defend itself against disease, pathogens and pests.

- Biopesticides may include natural plant-derived products, which include alkaloids, terpenoids, phenolics and other secondary chemicals. Certain vegetable oils such as canola oil are known to have pesticidal properties. Products based on extracts of plants such as garlic have now been registered in the EU and elsewhere.

Applications

Biopesticides are biological or biologically-derived agents, that are usually applied in a manner similar to chemical pesticides, but achieve pest management in an environmentally friendly way. With all pest management products, but especially microbial agents, effective control requires appropriate formulation and application.

Biopesticides for use against crop diseases have already established themselves on a variety of crops. For example, biopesticides already play an important role in controlling downy mildew diseases. Their benefits include: a 0-Day Pre-Harvest Interval, the ability to use under moderate to severe disease pressure, and the ability to use as a tank mix or in a rotational program with other registered fungicides. Because some market studies estimate that as much as 20% of global fungicide sales are directed at downy mildew diseases, the integration of biofungicides into grape production has substantial benefits in terms of extending the useful life of other fungicides, especially those in the reduced-risk category.

A major growth area for biopesticides is in the area of seed treatments and soil amendments. Fungicidal and biofungicidal seed treatments are used to control soil borne fungal pathogens that cause seed rots, damping-off, root rot and seedling blights. They can also be used to control internal seed–borne fungal pathogens as well as fungal pathogens that are on the surface of the seed. Many biofungicidal products also show capacities to stimulate plant host defence and other physiological processes that can make treated crops more resistant to a variety of biotic and abiotic stresses.

Advantages

- No harmful residues produced, i.e. biodegradable.

- Can be cheaper than chemical pesticides when locally produced.

- Can be more effective than synthetic pesticides in the long-term (as demonstrated, for example, by the LUBILOSA Programme)

Disadvantages

- High specificity: which may require an exact identification of the pest/pathogen and the use of multiple products to be used; although this can also be an advantage in that the biopesticide is less likely to harm species other than the target

- Often slow speed of action (thus making them unsuitable if a pest outbreak is an immediate threat to a crop)

- Often variable efficacy due to the influences of various biotic and abiotic factors (since some biopesticides are living organisms, which bring about pest/pathogen control by multiplying within or nearby the target pest/pathogen)

- Living organisms evolve and increase their resistance to biological, chemical, physical or any other form of control. If the target population is not exterminated or rendered incapable of reproduction, the surviving population can acquire a tolerance of whatever pressures are brought to bear, resulting in an evolutionary arms race.

- Unintended consequences: Studies have found broad spectrum biopesticides have lethal and nonlethal risks for non-target native pollinators such as Melipona quadrifasciata in Brazil.

Biological Pest Control

Syrphus hoverfly larva feeding on aphids

Parasitic wasp *Cotesia congregata* on tobacco hornworm *Manduca sexta*

Biological control is a method of controlling pests such as insects, mites, weeds and plant diseases using other organisms. It relies on predation, parasitism, herbivory, or other natural mechanisms, but typically also involves an active human management role. It can be an important component of integrated pest management (IPM) programs.

There are three basic types of biological pest control strategies: importation (sometimes called classical biological control), in which a natural enemy of a pest is introduced in the hope of achieving control; augmentation, in which locally-occurring natural enemies are bred and released to improve control; and conservation, in which measures are taken to increase natural enemies, such as by planting nectar-producing crop plants in the borders of rice fields.

Natural enemies of insect pests, also known as biological control agents, include predators, parasitoids, and pathogens. Biological control agents of plant diseases are most often referred to as antagonists. Biological control agents of weeds include seed predators, herbivores and plant pathogens.

Biological control can have side-effects on biodiversity through predation, parasitism, pathogenicity, competition, or other attacks on non-target species, especially when a species is introduced without thorough understanding of the possible consequences.

History

The term "biological control" was first used by Harry Scott Smith at the 1919 meeting of the Pacific Slope Branch of the American Association of Economic Entomologists, at the Mission Inn in downtown Riverside, California, and later defined by P. DeBach and K. S. Hagen in 1964. However, the practice has previously been used for centuries. The first report of the use of an insect species to control an insect pest comes from "Nan Fang Cao Mu Zhuang" (南方草木狀 *Plants of the Southern Regions*) (ca. 304 AD), which is attributed to Western Jin dynasty botanist *Ji Han* (嵇含, 263-307), in which it is mentioned that "*Jiaozhi people sell ants and their nests attached to twigs looking like thin cotton envelopes, the reddish-yellow ant being larger than normal. Without such ants, southern citrus fruits will be severely insect-damaged*". The ants used are known as *huang gan* (*huang* = yellow, *gan* = citrus) ants (*Oecophylla smaragdina*). The practice was later reported by Ling Biao Lu Yi (late Tang Dynasty or Early Five Dynasties), in *Ji Le Pian* by *Zhuang Jisu* (Southern Song Dynasty), in the *Book of Tree Planting* by Yu Zhen Mu (Ming Dynasty), in the book *Guangdong Xing Yu* (17th century), *Lingnan* by Wu Zhen Fang (Qing Dynasty), in *Nanyue Miscellanies* by Li Diao Yuan, and others.

Biological control techniques as we know them today started to emerge in the 1870s. During this decade, in the USA, the Missouri State Entomologist C. V. Riley and the Illinois State Entomologist W. LeBaron began within-state redistribution of parasitoids to control crop pests. The first international shipment of an insect as biological control agent was made by Charles V. Riley in 1873, shipping to France the predatory mites *Tyroglyphus phylloxera* to help fight the grapevine phylloxera (*Daktulosphaira vitifoliae*) that was destroying grapevines in France. The United States Department of Agriculture (USDA) initiated research in classical biological control following the establishment of the Division of Entomology in 1881, with C. V. Riley as Chief. The first importation of a parasitoid into the United States was this of *Cotesia glomerata* in 1883-1884, imported from Europe to control the imported cabbage white butterfly, *Pieris rapae*. In 1888-1889 the vedalia beetle, *Rodolia cardinalis*, which is a ladybug, was introduced from Australia to California to control the cottony cushion scale, *Icerya purchasi*. This had become a major problem for the newly developed citrus industry in California, and by the end of 1889 the cottony cushion scale population had already declined. This great success led to further introductions of beneficial insects into the USA.

In 1905 the USDA initiated its first large-scale biological control program, sending entomologists to Europe and Japan to look for natural enemies of the gypsy moth, *Lymantria dispar dispar*, and brown-tail moth, *Euproctis chrysorrhoea*, invasive pests of trees and shrubs. As a result, nine parasitoids of gypsy moth, seven of brown-tail moth, and two predators for both moths became established in the USA. Although the gypsy moth was not fully controlled by these natural enemies, the frequency, duration, and severity of its outbreaks were reduced and the program was regarded as successful. This program also led to the development of many concepts, principles, and procedures for the implementation of biological control programs.

The first reported case of a classical biological control attempt in Canada involves the hymenopteran parasitoid *Trichogramma minutum*. Individuals were caught in New York State and released in Ontario gardens in 1882 by William Saunders, trained chemist and first Director of the Dominion Experimental Farms, for controlling the imported currantworm *Nematus ribesii*. Between 1884 and 1908, the first Dominion Entomologist, James Fletcher, continued introductions of other parasitoids and pathogens for the control of pests in Canada.

Types of Biological Pest Control

There are three basic biological pest control strategies: importation (classical biological control), augmentation and conservation.

Importation

Rodolia cardinalis, the vedalia beetle, was imported to Australia in the 19th century, successfully controlling cottony cushion scale.

Importation or classical biological control involves the introduction of a pest's natural enemies to a new locale where they do not occur naturally. Early instances were often unofficial and not based on research, and some introduced species became serious pests themselves.

To be most effective at controlling a pest, a biological control agent requires a colonizing ability which allows it to keep pace with the spatial and temporal disruption of the habitat. Control is greatest if the agent has temporal persistence, so that it can maintain its population even in the temporary absence of the target species, and if it is an opportunistic forager, enabling it to rapidly exploit a pest population.

Joseph Needham noted a Chinese text dating from 304 AD, *Records of the Plants and Trees of the Southern Regions,* by Hsi Han, which describes mandarin oranges protected by large reddish-yellow citrus ants which attack and kill insect pests of the orange trees. The citrus ant (*Oecophylla smaragdina*) was rediscovered in the 20th century, and since 1958 has been used in China to protect orange groves.

One of the earliest successes in the west was in controlling *Icerya purchasi* (cottony cushion scale) in Australia, using a predatory insect *Rodolia cardinalis* (the vedalia beetle). This success was repeated in California using the beetle and a parasitoid fly, *Cryptochaetum iceryae.*

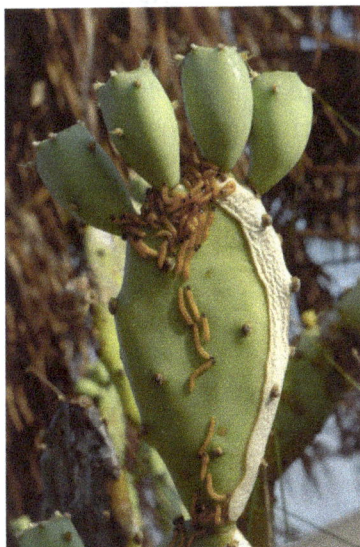

Cactoblastis cactorum larvae feeding on *Opuntia* cacti

Prickly pear cacti were introduced into Queensland, Australia as ornamental plants. They quickly spread to cover over 25 million hectares of Australia. Two control agents were used to help control the spread of the plant, the cactus moth *Cactoblastis cactorum,* and *Dactylopius* scale insects.

Damage from *Hypera postica,* the alfalfa weevil, a serious introduced pest of forage, was substantially reduced by the introduction of natural enemies. 20 years after their introduction the population of weevils in the alfalfa area treated for alfalfa weevil in the Northeastern United States remained 75 percent down.

The invasive species *Alternanthera philoxeroides* (alligator weed) was controlled in
Florida (U.S.) by introducing alligator weed flea beetle.

Alligator weed was introduced to the United States from South America. It takes root
in shallow water, interfering with navigation, irrigation, and flood control. The alligator
weed flea beetle and two other biological controls were released in Florida, enabling the
state to ban the use of herbicides to control alligator weed three years later. Another
aquatic weed, the giant salvinia (*Salvinia molesta*) is a serious pest, covering waterways, reducing water flow and harming native species. Control with the salvinia weevil
(*Cyrtobagous salviniae*) is effective in warm climates, and in Zimbabwe, a 99% control
of the weed was obtained over a two-year period.

Small commercially reared parasitoidal wasps, *Trichogramma ostriniae*, provide limited and erratic control of the European corn borer (*Ostrinia nubilalis*), a serious pest.
Careful formulations of the bacterium *Bacillus thuringiensis* are more effective.

The population of *Levuana iridescens*, the Levuana moth, a serious coconut pest in Fiji,
was brought under control by a classical biological control program in the 1920s.

Augmentation

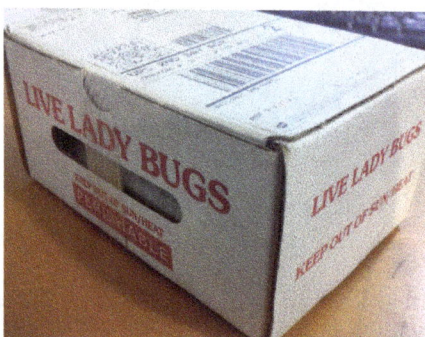

Hippodamia convergens, the convergent lady beetle, is commonly sold for biological
control of aphids.

Augmentation involves the supplemental release of natural enemies, boosting the naturally occurring population. In inoculative release, small numbers of the control agents
are released at intervals to allow them to reproduce, in the hope of setting up lon-
ger-term control, and thus keeping the pest down to a low level, constituting prevention rather than cure. In inundative release, in contrast, large numbers are released in

the hope of rapidly reducing a damaging pest population, correcting a problem that has already arisen. Augmentation can be effective, but is not guaranteed to work, and relies on understanding of the situation.

An example of inoculative release occurs in greenhouse production of several crops. Periodic releases of the parasitoid, *Encarsia formosa*, are used to control greenhouse whitefly, while the predatory mite *Phytoseiulus persimilis* is used for control of the two-spotted spider mite.

The egg parasite *Trichogramma* is frequently released inundatively to control harmful moths. Similarly, *Bacillus thuringiensis* and other microbial insecticides are similarly used in large enough quantities for a rapid effect. Recommended release rates for *Trichogramma* in vegetable or field crops range from 5,000 to 200,000 per acre (1 to 50 per square metre) per week according to the level of pest infestation. Similarly, entomopathogenic nematodes are released at rates of millions and even billions per acre for control of certain soil-dwelling insect pests.

Conservation

The conservation of existing natural enemies in an environment is the third method of biological pest control. Natural enemies are already adapted to the habitat and to the target pest, and their conservation can be simple and cost-effective, as when nectar-producing crop plants are grown in the borders of rice fields. These provide nectar to support parasitoids and predators of planthopper pests and have been demonstrated to be so effective (reducing pest densities by 10- or even 100-fold) that farmers sprayed 70% less insecticides, enjoyed yields boosted by 5%, and this led to an economic advantage of 7.5%. Predators of aphids were similarly found to be present in tussock grasses by field boundary hedges in England, but they spread too slowly to reach the centres of fields. Control was improved by planting a metre-wide strip of tussock grasses in field centres, enabling aphid predators to overwinter there.

An inverted flowerpot filled with straw to attract earwigs

Cropping systems can be modified to favor natural enemies, a practice sometimes referred to as habitat manipulation. Providing a suitable habitat, such as a shelterbelt, hedgerow, or beetle bank where beneficial insects can live and reproduce, can help ensure the survival of populations of natural enemies. Things as simple as leaving a layer of fallen leaves or mulch in place provides a suitable food source for worms and provides a shelter for insects, in turn being a food source for such beneficial mammals as hedgehogs and shrews. Compost piles and stacks of wood can provide shelter for invertebrates and small mammals. Long grass and ponds support amphibians. Not removing dead annuals and non-hardy plants in the autumn allows insects to make use of their hollow stems during winter. In California, prune trees are sometimes planted in grape vineyards to provide an improved overwintering habitat or refuge for a key grape pest parasitoid. The providing of artificial shelters in the form of wooden caskets, boxes or flowerpots is also sometimes undertaken, particularly in gardens, to make a cropped area more attractive to natural enemies. For example, earwigs are natural predators which can be encouraged in gardens by hanging upside-down flowerpots filled with straw or wood wool. Green lacewings can be encouraged by using plastic bottles with an open bottom and a roll of cardboard inside. Birdhouses enable insectivorous birds to nest; the most useful birds can be attracted by choosing an opening just large enough for the desired species.

Biological Control Agents

Predators

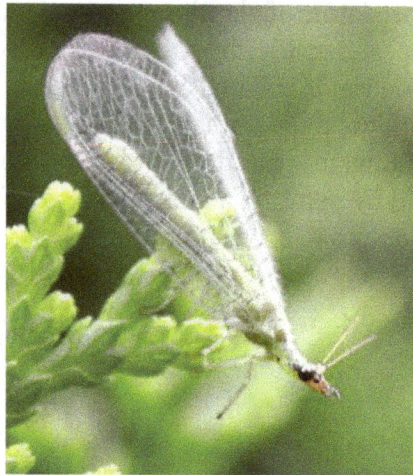

Lacewings are available from biocontrol dealers.

Predators are mainly free-living species that directly consume a large number of prey during their whole lifetime. Ladybugs, and in particular their larvae which are active between May and July in the northern hemisphere, are voracious predators of aphids, and also consume mites, scale insects and small caterpillars. The spotted lady beetle (*Coleomegilla maculata*) is also able to feed on the eggs and larvae of the Colorado potato beetle (*Leptinotarsa decemlineata*).

The larvae of many hoverfly species principally feed upon greenfly (aphids), one larva devouring up to 400 in its lifetime. Their effectiveness in commercial crops has not been studied.

Several species of entomopathogenic nematode are important predators of insect and other invertebrate pests. *Phasmarhabditis hermaphrodita* is a microscopic nematode that kills slugs. Its complex life cycle include a free-living, infective stage in the soil where it becomes associated with a pathogenic bacteria such as *Moraxella osloensis*. The nematode enters the slug through the posterior mantle region, thereafter feeding and reproducing inside, but it is the bacteria that kill the slug. The nematode is available commercially in Europe and is applied by watering onto moist soil.

Predatory *Polistes* wasp looking for bollworms or other caterpillars on a cotton plant

Species used to control spider mites include the predatory mites *Phytoseiulus persimilis, Neoseilus californicus,* and *Amblyseius cucumeris*, the predatory midge *Feltiella acarisuga*, and a ladybird *Stethorus punctillum*. The bug *Orius insidiosus* has been successfully used against the two-spotted spider mite and the western flower thrips (*Frankliniella occidentalis*).

Parasitoids

Parasitoids lay their eggs on or in the body of an insect host, which is then used as a food for developing larvae. The host is ultimately killed. Most insect parasitoids are wasps or flies, and may have a very narrow host range. The most important groups are the ichneumonid wasps, which prey mainly on caterpillars of butterflies and moths; braconid wasps, which attack caterpillars and a wide range of other insects including greenfly; chalcid wasps, which parasitize eggs and larvae of greenfly, whitefly, cabbage caterpillars, and scale insects; and tachinid flies, which parasitize a wide range of insects including caterpillars, adult and larval beetles, and true bugs.

Encarsia formosa was one of the first biological control agents developed.

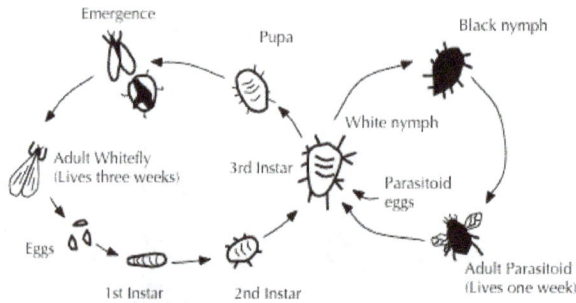

Life cycles of Greenhouse whitefly and its parasitoid wasp *Encarsia formosa*

Encarsia formosa is a small predatory chalcid wasp which is a parasitoid of white-fly, a sap-feeding insect which can cause wilting and black sooty moulds in glasshouse vegetable and ornamental crops. It is most effective when dealing with low level infestations, giving protection over a long period of time. The wasp lays its eggs in young whitefly 'scales', turning them black as the parasite larvae pupates. *Gonatocerus ashmeadi* (Hymenoptera: Mymaridae) has been introduced to control the glassy-winged sharpshooter *Homalodisca vitripennis* (Hemipterae: Cicadellidae) in French Polynesia and has successfully controlled ~95% of the pest density.

Parasitoids are among the most widely used biological control agents. Commercially, there are two types of rearing systems: short-term daily output with high production of parasitoids per day, and long-term low daily output with a range in production of 4-1000million female parasitoids per week. Larger production facilities produce on a yearlong basis, whereas some facilities produce only seasonally. Rearing facilities are usually a significant distance from where the agents are to be used in the field, and transporting the parasitoids from the point of production to the point of use can pose problems. Shipping conditions can be too hot, and even vibrations from planes or trucks can adversely affect parasitoids.

Pathogens

Pathogenic micro-organisms include bacteria, fungi, and viruses. They kill or debilitate their host and are relatively host-specific. Various microbial insect diseases occur naturally, but may also be used as biological pesticides. When naturally occurring, these outbreaks are density-dependent in that they generally only occur as insect populations become denser.

Bacteria

Bacteria used for biological control infect insects via their digestive tracts, so they offer only limited options for controlling insects with sucking mouth parts such as aphids and scale insects. *Bacillus thuringiensis* is the most widely applied species of bacteria used for biological control, with at least four sub-species used against Lepidopteran (moth, butterfly), Coleopteran (beetle) and Dipteran (true fly) insect pests. The bacterium is available in sachets of dried spores which are mixed with water and sprayed onto vulnerable plants such as brassicas and fruit trees. *B. thuringiensis* has also been incorporated into crops, making them resistant to these pests and thus reducing the use of pesticides. The bacterium *Paenibacillus popilliae* causes milky spore disease has been found useful in the control of Japanese beetle, killing the larvae. It is very specific to its host species and is harmless to vertebrates and other invertebrates.

Fungi

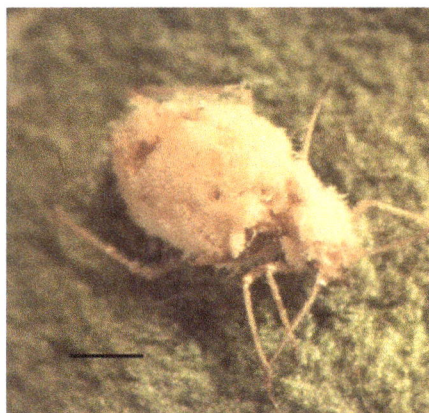

Green peach aphid, a pest in its own right and a vector of plant viruses, killed by the fungus *Pandora neoaphidis* (Zygomycota: Entomophthorales) Scale bar = 0.3 mm.

Entomopathogenic fungi, which cause disease in insects, include at least 14 species that attack aphids. *Beauveria bassiana* is mass-produced and used to manage a wide variety of insect pests including whiteflies, thrips, aphids and weevils. *Lecanicillium* spp. are deployed against white flies, thrips and aphids. *Metarhizium* spp. are used against pests including beetles, locusts and other grasshoppers, Hemiptera, and spider mites. *Paecilomyces fumosoroseus* is effective against white flies, thrips and aphids; *Purpureocillium lilacinus* is used against root-knot nematodes, and 89 *Trichoderma* species against certain plant pathogens. *Trichoderma viride* has been used against Dutch elm disease, and has shown some effect in suppressing silver leaf, a disease of stone fruits caused by the pathogenic fungus *Chondrostereum purpureum*.

The fungi *Cordyceps* and *Metacordyceps* are deployed against a wide spectrum of arthropods. *Entomophaga* is effective against pests such as the green peach aphid.

Several members of Chytridiomycota and Blastocladiomycota have been explored as agents of biological control. From Chytridiomycota, *Synchytrium solstitiale* is being considered as a control agent of the yellow star thistle (*Centaurea solstitialis*) in the United States.

Viruses

Baculoviruses are specific to individual insect host species and have been shown to be useful in biological pest control. For example, the Lymantria dispar multicapsid nuclear polyhedrosis virus has been used to spray large areas of forest in North America where larvae of the gypsy moth are causing serious defoliation. The moth larvae are killed by the virus they have eaten and die, the disintegrating cadavers leaving virus particles on the foliage to infect other larvae.

A mammalian virus, the rabbit haemorrhagic disease virus has been introduced to Australia and to New Zealand to attempt to control the European rabbit populations there.

Algae

Lagenidium giganteum is a water-borne mould that parasitizes the larval stage of mosquitoes. When applied to water, the motile spores avoid unsuitable host species and search out suitable mosquito larval hosts. This alga has the advantages of a dormant phase, resistant to desiccation, with slow-release characteristics over several years. Unfortunately, it is susceptible to many chemicals used in mosquito abatement programmes.

Plants

The legume vine *Mucuna pruriens* is used in the countries of Benin and Vietnam as a biological control for problematic *Imperata cylindrica* grass. *Mucuna pruriens* is said not to be invasive outside its cultivated area. *Desmodium uncinatum* can be used in push-pull farming to stop the parasitic plant, *Striga*.

Other Methods

Combined use of Parasitoids and Pathogens

In cases of massive and severe infection of invasive pests, techniques of pest control are often used in combination. An example is the emerald ash borer, *Agrilus planipennis*, an invasive beetle from China, which has destroyed tens of millions of ash trees in its introduced range in North America. As part of the campaign against it, from 2003 American scientists and the Chinese Academy of Forestry searched for its natural enemies in the wild, leading to the discovery of several parasitoid wasps, namely *Tet-*

rastichus planipennisi, a gregarious larval endoparasitoid,*Oobius agrili*, a solitary, parthenogenic egg parasitoid, and *Spathius agrili*, a gregarious larval ectoparasitoid. These have been introduced and released into the United States of America as a possible biological control of the emerald ash borer. Initial results have shown promise with *Tetrastichus planipennisi* and it is now being released along with *Beauveria bassiana*, a fungal pathogen with known insecticidal properties.

Indirect Control

Pests may be controlled by biological control agents that do not prey directly upon them. For example, the Australian bush fly, *Musca vetustissima*, is a major nuisance pest in Australia, but native decomposers found in Australia are not adapted to feeding on cow dung, which is where bush flies breed. Therefore, the Australian Dung Beetle Project (1965–1985), led by Dr. George Bornemissza of the Commonwealth Scientific and Industrial Research Organisation, released forty-nine species of dung beetle, with the aim of reducing the amount of dung and therefore also the potential breeding sites of the fly.

Side-effects

Biological control can affect biodiversity through predation, parasitism, pathogenicity, competition, or other attacks on non-target species. An introduced control does not always target only the intended pest species; it can also target native species. In Hawaii during the 1940s parasitic wasps were introduced to control a lepidopteran pest and the wasps are still found there today. This may have a negative impact on the native ecosystem, however, host range and impacts need to be studied before declaring their impact on the environment.

Vertebrate animals tend to be generalist feeders, and seldom make good biological control agents; many of the classic cases of "biocontrol gone awry" involve vertebrates. For example, the cane toad (*Bufo marinus*) was intentionally introduced to Australia to control the greyback cane beetle (*Dermolepida albohirtum*), and other pests of sugar cane. 102 toads were obtained from Hawaii and bred in captivity to increase their numbers until they were released into the sugar cane fields of the tropic north in 1935. It was later discovered that the toads could not jump very high and so were unable to eat the cane beetles which stayed up on the upper stalks of the cane plants. However the toad thrived by feeding on other insects and it soon spread very rapidly; it took over native amphibian habitat and brought foreign disease to native toads and frogs, dramatically reducing their populations. Also when it is threatened or handled, the cane toad releases poison from parotoid glands on its shoulders; native Australian species such as goannas, tiger snakes, dingos and northern quolls that attempted to eat the toad were harmed or killed. However, there has been some recent evidence that native predators are adapting, both physiologically and through changing their behaviour, so in the long run, their populations may recover.

Rhinocyllus conicus, a seed-feeding weevil, was introduced to North America to control exotic musk thistle (*Carduus nutans*) and Canadian thistle (*Cirsium arvense*). However the weevil also attacks native thistles, harming such species as the endemic Platte thistle (*Cirsium neomexicanum*) by selecting larger plants (which reduced the gene pool), reducing seed production and ultimately threatening the species' survival.

The small Asian mongoose (*Herpestus javanicus*) was introduced to Hawaii in order to control the rat population. However it was diurnal and the rats emerged at night, and it preyed on the endemic birds of Hawaii, especially their eggs, more often than it ate the rats, and now both rats and mongooses threaten the birds. This introduction was undertaken without understanding the consequences of such an action. No regulations existed at the time, and more careful evaluation should prevent such releases now.

The sturdy and prolific eastern mosquitofish (*Gambusia holbrooki*) is a native of the southeastern United States and was introduced around the world in the 1930s and 40s to feed on mosquito larvae and thus combat malaria. However, it has thrived at the expense of local species, causing a decline of endemic fish and frogs through competition for food resources, as well as through eating their eggs and larvae. In Australia, the mosquitofish is the subject of discussion as to how best to control it; in 1989 it was said that "biological population control is well beyond present capabilities", and this remains the position.

Grower Education

A potential obstacle to the adoption of biological pest control measures is growers sticking to the familiar use of pesticides. It has been claimed that many of the pests that are controlled today using pesticides, actually became pests because pesticide use reduced or eliminated natural predators. A method of increasing grower adoption of biocontrol involves is letting growers learn by doing, for example showing them simple field experiments, having observations of live predation of pests, or collections of parasitised pests. In the Philippines, early season sprays against leaf folder caterpillars were common practice, but growers were asked to follow a 'rule of thumb' of not spraying against leaf folders for the first 30 days after transplanting; participation in this resulted in a reduction of insecticide use by 1/3 and a change in grower perception of insecticide use.

References

- Fred Baur. Insect Management for Food Storage and Processing. American Association of Cereal Chemists. ISBN 0-913250-38-4.

- Charles Perrings; Mark Herbert Williamson; Silvana Dalmazzone (1 January 2000). The Economics of Biological Invasions. Edward Elgar Publishing. ISBN 978-1-84064-378-7.

- Consoli, Fernando L.; Parra, José Roberto Postali; Trichogramma, Roberto Antônio Zucchi (28 September 2010). Egg Parasitoids in Agroecosystems with Emphasis on. Springer. ISBN 978-1-4020-9110-0.

- Rajeev K. Upadhyay; K.G. Mukerji; B. P. Chamola (30 November 2001). Biocontrol Potential and its Exploitation in Sustainable Agriculture: Volume 2: Insect Pests. Springer. pp. 261–. ISBN 978-0-306-46587-1.

- J. C. van Lenteren (2003). Quality Control and Production of Biological Control Agents: Theory and Testing Procedures. CABI. ISBN 978-0-85199-836-7.

- Copping, Leonard G. (2009). The Manual of Biocontrol Agents: A World Compendium. BCPC. ISBN 978-1-901396-17-1.

- Francis Borgio J, Sahayaraj K and Alper Susurluk I (eds) . Microbial Insecticides: Principles and Applications, Nova Publishers, USA. 492pp. ISBN 978-1-61209-223-2

- Flint, Maria Louise & Dreistadt, Steve H. (1998). Clark, Jack K., ed. Natural Enemies Handbook: The Illustrated Guide to Biological Pest Control. University of California Press. ISBN 978-0-520-21801-7.

- Acorn, John (2007). Ladybugs of Alberta: Finding the Spots and Connecting the Dots. University of Alberta. p. 15. ISBN 978-0-88864-381-0.

- Kaya, Harry K. et al. (1993). "An Overview of Insect-Parasitic and Entomopathogenic Nematodes". In Bedding, R.A. Nematodes and the Biological Control of Insect Pests. CSIRO Publishing. ISBN 978-0-643-10591-1.

- "Classical Biological Control: Importation of New Natural Enemies". University of Wisconsin. Retrieved 7 June 2016.

- "How to Manage Pests. Cottony Cushion Scale". University of California Integrated Pest Management. Retrieved 5 June 2016.

- Shapiro-Ilan, David I; Gaugler, Randy. "Biological Control. Nematodes (Rhabditida: Steinernematidae & Heterorhabditidae)". Cornell University. Retrieved 7 June 2016.

- "Conservation of Natural Enemies: Keeping Your "Livestock" Happy and Productive". University of Wisconsin. Retrieved 7 June 2016.

- Capinera, John L. (October 2005). "Featured creatures:". University of Florida website - Department of Entomology and Nematology. University of Florida. Retrieved 7 June 2016.

- "The cane toad (Bufo marinus)". Australian Government: Department of the Environment. 2010. Retrieved 2 July 2016.

- "Moving on from the mongoose: the success of biological control in Hawai'i". Kia'i Moku. MISC. 18 April 2012. Retrieved 2 July 2016.

- "With BioDirect, Monsanto Hopes RNA Sprays Can Someday Deliver Drought Tolerance and Other Traits to Plants on Demand | MIT Technology Review". Retrieved 2015-08-31.

- "Help WildCare Pursue Stricter Rodenticide Controls in California". wildcarebayarea.org/. Wild Care. Retrieved 28 February 2014.

- "Safer Rodenticide Products". epa.gov. USA Environment Protection Agency. March 2013. Retrieved 23 February 2014.

- WOODY, TODD (September 20, 2010). "A Crop Sprouts Without Soil or Sunshine". nytimes.com. The New York Times. Retrieved 28 February 2014.

3

Fertilizer: An Overview

Fertilizers are applied on plants in order to supply them with more nutrients. They help with the growth of plants and also improve the efficiency of the soil. The chemical compounds that are used in fertilizers are ammonium nitrate, calcium nitrate, potassium nitrate and monocalcium phosphate. This chapter is an overview of the subject matter incorporating all the major aspects of fertilizers.

Fertilizer

A large, modern fertilizer spreader

A fertilizer (American English) or fertiliser (British English) is any material of natural or synthetic origin (other than liming materials) that is applied to soils or to plant tissues (usually leaves) to supply one or more plant nutrients essential to the growth of plants.

Mechanism

Six tomato plants grown with and without nitrate fertilizer on nutrient-poor sand/clay soil. One of the plants in the nutrient-poor soil has died.

Fertilizers enhance the growth of plants. This goal is met in two ways, the traditional one being additives that provide nutrients. The second mode by some fertilizers act is to enhance the effectiveness of the soil by modifying its water retention and aeration. This article, like many on fertilizers, emphasises the nutritional aspect. Fertilizers typically provide, in varying proportions:

- three main macronutrients:

    o   Nitrogen (N): leaf growth;

    o   Phosphorus (P): Development of roots, flowers, seeds, fruit;

    o   Potassium (K): Strong stem growth, movement of water in plants, pro-
        motion of flowering and fruiting;

- three secondary macronutrients: calcium (Ca), magnesium (Mg), and sulphur
  (S);

- micronutrients: copper (Cu), iron (Fe), manganese (Mn), molybdenum (Mo),
  zinc (Zn), boron (B), and of occasional significance there are silicon (Si), cobalt
  (Co), and vanadium (V) plus rare mineral catalysts.

A Lite-Trac Agri-Spread lime and fertilizer spreader at an agricultural show

The nutrients required for healthy plant life are classified according to the elements,
but the elements are not used as fertilizers. Instead compounds containing these elements are the basis of fertilisers. The macronutrients are consumed in larger quantities
and are present in plant tissue in quantities from 0.15% to 6.0% on a dry matter (DM)
(0% moisture) basis. Plants are made up of four main elements: hydrogen, oxygen,
carbon, and nitrogen. Carbon, hydrogen and oxygen are widely available as water and
carbon dioxide. Although nitrogen makes up most of the atmosphere, it is in a form
that is unavailable to plants. Nitrogen is the most important fertilizer since nitrogen is
present in proteins, DNA and other components (e.g., chlorophyll). To be nutritious to
plants, nitrogen must be made available in a "fixed" form. Only some bacteria and their
host plants (notably legumes) can fix atmospheric nitrogen (N_2) by converting it to
ammonia. Phosphate is required for the production of DNA and ATP, the main energy
carrier in cells, as well as certain lipids.

Micronutrients are consumed in smaller quantities and are present in plant tissue on the order of parts-per-million (ppm), ranging from 0.15 to 400 ppm DM, or less than 0.04% DM. These elements are often present at the active sites of enzymes that carry out the plant's metabolism. Because these elements enable catalysts (enzymes) their impact far exceeds their weight percentage.

Classification

Fertilizers are classified in several ways. They are classified according to whether they provide a single nutrient (say, N, P, or K), in which case they are classified as "straight fertilizers." "Multinutrient fertilizers" (or "complex fertilizers") provide two or more nutrients, for example N and P. Fertilizers are also sometimes classified as inorganic versus organic. Inorganic fertilizers exclude carbon-containing materials except ureas. Organic fertilizers are usually (recycled) plant- or animal-derived matter. Inorganic are sometimes called synthetic fertilizers since various chemical treatments are required for their manufacture.

Single Nutrient ("Straight") Fertilizers

The main nitrogen-based straight fertilizer is ammonia or its solutions. Ammonium nitrate (NH_4NO_3) is also widely used. About 15M tons were produced in 1981. Urea is another popular source of nitrogen, having the advantage that it is a solid and non-explosive, unlike ammonia and ammonium nitrate, respectively. A few percent of the nitrogen fertilizer market (4% in 2007) has been met by calcium ammonium nitrate ($Ca(NO_3)_2 \cdot NH_4NO_3 \cdot 10H_2O$).

The main straight phosphate fertilizers are the superphosphates. "Single superphosphate" (SSP) consists of 14–18% P_2O_5, again in the form of $Ca(H_2PO_4)_2$, but also phosphogypsum ($CaSO_4 \cdot 2 H_2O$). Triple superphosphate (TSP) typically consists of 44-48% of P_2O_5 and no gypsum. A mixture of single superphosphate and triple superphosphate is called double superphosphate. More than 90% of a typical superphosphate fertilizer is water-soluble.

Multinutrient Fertilizers

These fertilizers are the most common. They consist of two or more nutrient components.

Binary (NP, NK, PK) fertilizers

Major two-component fertilizers provide both nitrogen and phosphorus to the plants. These are called NP fertilizers. The main NP fertilizers are monoammonium phosphate (MAP) and diammonium phosphate (DAP). The active ingredient in MAP is $NH_4H_2PO_4$. The active ingredient in DAP is $(NH_4)_2HPO_4$. About 85% of MAP and DAP fertilizers are soluble in water.

NPK Fertilizers

NPK fertilizers are three-component fertilizers providing nitrogen, phosphorus, and potassium.

NPK rating is a rating system describing the amount of nitrogen, phosphorus, and potassium in a fertilizer. NPK ratings consist of three numbers separated by dashes (e.g., 10-10-10 or 16-4-8) describing the chemical content of fertilizers. The first number represents the percentage of nitrogen in the product; the second number, P_2O_5; the third, K_2O. Fertilizers do not actually contain P_2O_5 or K_2O, but the system is a conventional shorthand for the amount of the phosphorus (P) or potassium (K) in a fertilizer. A 50-pound (23 kg) bag of fertilizer labeled 16-4-8 contains 8 lb (3.6 kg) of nitrogen (16% of the 50 pounds), an amount of phosphorus equivalent to that in 2 pounds of P_2O_5 (4% of 50 pounds), and 4 pounds of K_2O (8% of 50 pounds). Most fertilizers are labeled according to this N-P-K convention, although Australian convention, following an N-P-K-S system, adds a fourth number for sulfur.

Micronutrients

The main micronutrients are molybdenum, zinc, and copper. These elements are provided as water-soluble salts. Iron presents special problems because it converts to insoluble (bio-unavailable) compounds at moderate soil pH and phosphate concentrations. For this reason, iron is often administered as a chelate complex, e.g., the EDTA derivative. The micronutrient needs depend on the plant. For example, sugar beets appear to require boron, and legumes require cobalt.

Production

Nitrogen Fertilizers

Top users of Nitrogen-based Fertilizer		
Country	Total N use (Mt pa)	Amt. used for feed/ pasture (Mt pa)
China	18.7	3.0
India	11.9	N/A
U.S.	9.1	4.7
France	2.5	1.3
Germany	2.0	1.2
Brazil	1.7	0.7
Canada	1.6	0.9

Turkey	1.5	0.3
UK	1.3	0.9
Mexico	1.3	0.3
Spain	1.2	0.5
Argentina	0.4	0.1

Nitrogen fertilizers are made from ammonia (NH_3), which is sometimes injected into the ground directly. The ammonia is produced by the Haber-Bosch process. In this energy-intensive process, natural gas (CH_4) supplies the hydrogen, and the nitrogen (N_2) is derived from the air. This ammonia is used as a feedstock for all other nitrogen fertilizers, such as anhydrous ammonium nitrate (NH_4NO_3) and urea ($CO(NH_2)_2$).

Deposits of sodium nitrate ($NaNO_3$) (Chilean saltpeter) are also found in the Atacama desert in Chile and was one of the original (1830) nitrogen-rich fertilizers used. It is still mined for fertilizer.

There has been technical work investigating on-site (on-farm) synthesis of nitrate fertilizer using solar photovoltaic power, which would enable farmers more control in soil fertility, while using far less surface area than conventional organic farming for nitrogen fertilizer.

Phosphate Fertilizers

All phosphate fertilizers are obtained by extraction from minerals containing the anion PO_4^{3-}. In rare cases, fields are treated with the crushed mineral, but most often more soluble salts are produced by chemical treatment of phosphate minerals. The most popular phosphate-containing minerals are referred to collectively as phosphate rock. The main minerals are fluorapatite $Ca_5(PO_4)_3F$ (CFA) and hydroxyapatite $Ca_5(PO_4)_3OH$. These minerals are converted to water-soluble phosphate salts by treatment with sulfuric or phosphoric acids. The large production of sulfuric acid as an industrial chemical is primarily due to its use as cheap acid in processing phosphate rock into phosphate fertilizer. The global primary uses for both sulfur and phosphorus compounds relate to this basic process.

In the nitrophosphate process or Odda process (invented in 1927), phosphate rock with up to a 20% phosphorus (P) content is dissolved with nitric acid (HNO_3) to produce a mixture of phosphoric acid (H_3PO_4) and calcium nitrate ($Ca(NO_3)_2$). This mixture can be combined with a potassium fertilizer to produce a *compound fertilizer* with the three macronutrients N, P and K in easily dissolved form.

Potassium Fertilizers

Potash is a mixture of potassium minerals used to make potassium (chemical symbol: K) fertilizers. Potash is soluble in water, so the main effort in producing this nutrient from the ore involves some purification steps; e.g., to remove sodium chloride (NaCl) (common salt). Sometimes potash is referred to as K_2O, as a matter of convenience to those describing the potassium content. In fact potash fertilizers are usually potassium chloride, potassium sulfate, potassium carbonate, or potassium nitrate.

Compound Fertilizers

Compound fertilizers, which contain N, P, and K, can often be produced by mixing straight fertilizers. In some cases, chemical reactions occur between the two or more components. For example, monoammonium and diammonium phosphates, which provide plants with both N and P, are produced by neutralizing phosphoric acid (from phosphate rock) and ammonia :

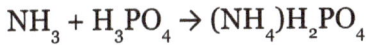

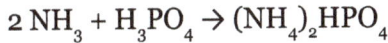

$$NH_3 + H_3PO_4 \rightarrow (NH_4)H_2PO_4$$

$$2\,NH_3 + H_3PO_4 \rightarrow (NH_4)_2HPO_4$$

Organic Fertilizers

Compost bin for small-scale production of organic fertilizer

A large commercial compost operation

The main "organic fertilizers" are peat, animal wastes, plant wastes from agriculture, and treated sewage sludge (biosolids). In terms of volume, peat is the most widely used organic fertilizer. This immature form of coal confers no nutritional value to the plants, but improves the soil by aeration and absorbing water. Animal sources include the products of the slaughter of animals. Bloodmeal, bone meal, hides, hoofs, and horns are typical components. Organic fertilizer usually contain fewer nutrients, but offer other advantages as well as being appealing to those who are trying to practice "environmentally friendly" farming.

Other Elements: Calcium, Magnesium, and Sulfur

Calcium is supplied as superphosphate or calcium ammonium nitrate solutions.

Application

Fertilizers are commonly used for growing all crops, with application rates depending on the soil fertility, usually as measured by a soil test and according to the particular crop. Legumes, for example, fix nitrogen from the atmosphere and generally do not require nitrogen fertilizer.

Liquid Vs Solid

Fertilizers are applied to crops both as solids and as liquid. About 90% of fertilizers are applied as solids. Solid fertilizer is typically granulated or powdered. Often solids are available as prills, a solid globule. Liquid fertilizers comprise anhydrous ammonia, aqueous solutions of ammonia, aqueous solutions of ammonium nitrate or urea. These concentrated products may be diluted with water to form a concentrated liquid fertilizer (e.g., UAN). Advantages of liquid fertilizer are its more rapid effect and easier coverage. The addition of fertilizer to irrigation water is called "fertigation".

Slow- and Controlled-release Fertilizers

Slow- and controlled-release involve only 0.15% (562,000 tons) of the fertilizer market (1995). Their utility stems from the fact that fertilizers are subject to antagonistic processes. In addition to their providing the nutrition to plants, excess fertilizers can be poisonous to the same plant. Competitive with the uptake by plants is the degradation or loss of the fertilizer. Microbes degrade many fertilizers, e.g., by immobilization or oxidation. Furthermore, fertilizers are lost by evaporation or leaching. Most slow-release fertilizers are derivatives of urea, a straight fertilizer providing nitrogen. Isobutylidenediurea ("IBDU") and urea-formaldehyde slowly convert in the soil to free urea, which is rapidly uptaken by plants. IBDU is a single compound with the formula $(CH_3)_2CHCH(NHC(O)NH_2)_2$ whereas the urea-formaldehydes consist of mixtures of the approximate formula $(HOCH_2NHC(O)NH)_nCH_2$.

Besides being more efficient in the utilization of the applied nutrients, slow-release technologies also reduce the impact on the environment and the contamination of the subsurface water. Slow-release fertilizers (various forms including fertilizer spikes, tabs, etc.) which reduce the problem of "burning" the plants due to excess nitrogen.

Polymer coating of fertilizer ingredients gives tablets and spikes a 'true time-release' or 'staged nutrient release' (SNR) of fertilizer nutrients.

Controlled release fertilizers are traditional fertilizers encapsulated in a shell that degrades at a specified rate. Sulfur is a typical encapsulation material. Other coated products use thermoplastics (and sometimes ethylene-vinyl acetate and surfactants, etc.) to produce diffusion-controlled release of urea or other fertilizers. "Reactive Layer Coating" can produce thinner, hence cheaper, membrane coatings by applying reactive monomers simultaneously to the soluble particles. "Multicote" is a process applying layers of low-cost fatty acid salts with a paraffin topcoat.

Foliar Application

Foliar fertilizers are applied directly to leaves. The method is almost invariably used to apply water-soluble straight nitrogen fertilizers and used especially for high value crops such as fruits.

Fertilizer burn

Chemicals that Affect Nitrogen Uptake

Various chemicals are used to enhance the efficiency of nitrogen-based fertilizers. In this way farmers can limit the polluting effects of nitrogen run-off. Nitrification inhibitors (also known as nitrogen stabilizers) suppress the conversion of ammonia into nitrate, an anion that is more prone to leaching. 1-Carbamoyl-3-methylpyrazole (CMP), dicyandiamide, and nitrapyrin (2-chloro-6-trichloromethylpyridine) are popular. Urease inhibitors are used to slow the hydrolytic conversion of urea into ammonia, which is prone to evaporation as well as nitrification. The conversion of urea to ammonia catalyzed by enzymes called ureases. A popular inhibitor of ureases is N-(n-butyl)thiophosphoric triamide (NBPT).

Overfertilization

Careful fertilization technologies are important because excess nutrients can be as detrimental. Fertilizer burn can occur when too much fertilizer is applied, resulting in drying out of the leaves and damage or even death of the plant. Fertilizers vary in their tendency to burn roughly in accordance with their salt index.

Statistics

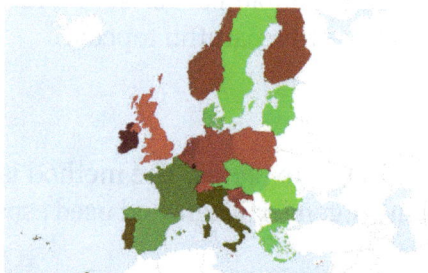

The map displays the statistics of fertilizer consumption in western and central European counties from data published by The World Bank for 2012.

Conservative estimates report 30 to 50% of crop yields are attributed to natural or synthetic commercial fertilizer. Global market value is likely to rise to more than US$185 billion until 2019. The European fertilizer market will grow to earn revenues of approx. €15.3 billion in 2018.

Data on the fertilizer consumption per hectare arable land in 2012 are published by The World Bank. For the diagram below values of the European Union (EU) countries have been extracted and are presented as kilograms per hectare (pounds per acre). The total consumption of fertilizer in the EU is 15.9 million tons for 105 million hectare arable land area (or 107 million hectare arable land according to another estimate). This figure equates to 151 kg of fertilizers consumed per ha arable land on average for the EU countries. Interestingly, mainly in those countries where fertilizers are consumed a lot also plant growth product are sold more than in others.

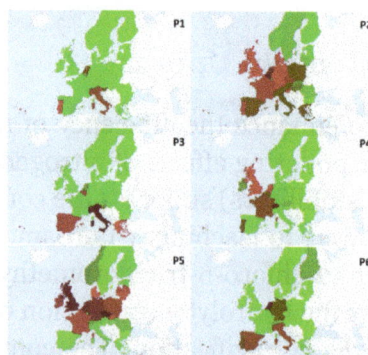

Pesticide categories, EUROSTAT. P5= Plant growth regulators. The red/green scale represents high/low pesticide sales per arable land.

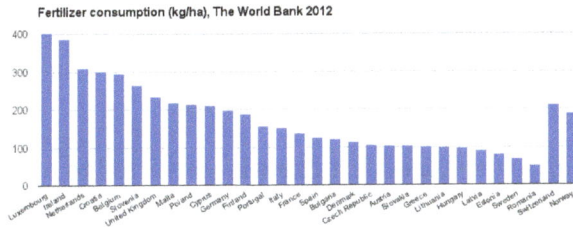
Fertilizer consumption (kg/ha), The World Bank 2012

Environmental Effects

Runoff of soil and fertilizer during a rain storm

An algal bloom caused by eutrophication

Water

Agricultural run-off is a major contributor to the eutrophication of fresh water bodies. For example, in the US, about half of all the lakes are eutrophic. The main contributor to eutrophication is phosphate, which is normally a limiting nutrient; high concentrations promote the growth of cyanobacteria and algae, the demise of which consumes oxygen. Cyanobacteria blooms ('algal blooms') can also produce harmful toxins that can accumulate in the food chain, and can be harmful to humans.

The nitrogen-rich compounds found in fertilizer runoff are the primary cause of serious oxygen depletion in many parts of oceans, especially in coastal zones, lakes and rivers. The resulting lack of dissolved oxygen greatly reduces the ability of these areas to sustain oceanic fauna. The number of oceanic dead zones near inhabited coastlines are increasing. As of 2006, the application of nitrogen fertilizer is being increasingly controlled in northwestern Europe and the United States. If eutrophication *can* be reversed, it may take decades before the accumulated nitrates in groundwater can be broken down by natural processes.

Nitrate Pollution

Only a fraction of the nitrogen-based fertilizers is converted to produce and other plant matter. The remainder accumulates in the soil or lost as run-off. High application rates of nitrogen-containing fertilizers combined with the high water solubility of nitrate leads to increased runoff into surface water as well as leaching into groundwater, thereby causing groundwater pollution. The excessive use of nitrogen-containing fertilizers (be they synthetic or natural) is particularly damaging, as much of the nitrogen that is not taken up by plants is transformed into nitrate which is easily leached.

Nitrate levels above 10 mg/L (10 ppm) in groundwater can cause 'blue baby syndrome' (acquired methemoglobinemia). The nutrients, especially nitrates, in fertilizers can cause problems for natural habitats and for human health if they are washed off soil into watercourses or leached through soil into groundwater.

Soil

Acidification

Nitrogen-containing fertilizers can cause soil acidification when added. This may lead to decreases in nutrient availability which may be offset by liming.

Accumulation of Toxic Elements

Cadmium

The concentration of cadmium in phosphorus-containing fertilizers varies considerably and can be problematic. For example, mono-ammonium phosphate fertilizer may have a cadmium content of as low as 0.14 mg/kg or as high as 50.9 mg/kg. This is because the phosphate rock used in their manufacture can contain as much as 188 mg/kg cadmium (examples are deposits on Nauru and the Christmas islands). Continuous use of high-cadmium fertilizer can contaminate soil (as shown in New Zealand) and plants. Limits to the cadmium content of phosphate fertilizers has been considered by the European Commission. Producers of phosphorus-containing fertilizers now select phosphate rock based on the cadmium content.

Fluoride

Phosphate rocks contain high levels of fluoride. Consequently, the widespread use of phosphate fertilizers has increased soil fluoride concentrations. It has been found that food contamination from fertilizer is of little concern as plants accumulate little fluoride from the soil; of greater concern is the possibility of fluoride toxicity to livestock that ingest contaminated soils. Also of possible concern are the effects of fluoride on soil microorganisms.

Radioactive Elements

The radioactive content of the fertilizers varies considerably and depends both on their concentrations in the parent mineral and on the fertilizer production process. Uranium-238 concentrations range can range from 7 to 100 pCi/g in phosphate rock and from 1 to 67 pCi/g in phosphate fertilizers. Where high annual rates of phosphorus fertilizer are used, this can result in uranium-238 concentrations in soils and drainage waters that are several times greater than are normally present. However, the impact of these increases on the risk to human health from radinuclide contamination of foods is very small (less than 0.05 mSv/y).

Other Metals

Steel industry wastes, recycled into fertilizers for their high levels of zinc (essential to plant growth), wastes can include the following toxic metals: lead arsenic, cadmium, chromium, and nickel. The most common toxic elements in this type of fertilizer are mercury, lead, and arsenic. These potentially harmful impurities can be removed; however, this significantly increases cost. Highly pure fertilizers are widely available and perhaps best known as the highly water-soluble fertilizers containing blue dyes used around households, such as Miracle-Gro. These highly water-soluble fertilizers are used in the plant nursery business and are available in larger packages at significantly less cost than retail quantities. There are also some inexpensive retail granular garden fertilizers made with high purity ingredients.

Trace Mineral Depletion

Attention has been addressed to the decreasing concentrations of elements such as iron, zinc, copper and magnesium in many foods over the last 50–60 years. Intensive farming practices, including the use of synthetic fertilizers are frequently suggested as reasons for these declines and organic farming is often suggested as a solution. Although improved crop yields resulting from NPK fertilizers are known to dilute the concentrations of other nutrients in plants, much of the measured decline can be attributed to the use of progressively higher-yielding crop varieties which produce foods with lower mineral concentrations than their less productive ancestors. It is, therefore, unlikely that organic farming or reduced use of fertilizers will solve the problem; foods with high nutrient density are posited to be achieved using older, lower-yielding varieties or the development of new high-yield, nutrient-dense varieties.

Fertilizers are, in fact, more likely to solve trace mineral deficiency problems than cause them: In Western Australia deficiencies of zinc, copper, manganese, iron and molybdenum were identified as limiting the growth of broad-acre crops and pastures in the

1940s and 1950s. Soils in Western Australia are very old, highly weathered and deficient in many of the major nutrients and trace elements. Since this time these trace elements are routinely added to fertilizers used in agriculture in this state. Many other soils around the world are deficient in zinc, leading to deficiency in both plants and humans, and zinc fertilizers are widely used to solve this problem.

Changes in Soil Biology

High levels of fertilizer may cause the breakdown of the symbiotic relationships between plant roots and mycorrhizal fungi.

Energy Consumption and Sustainability

In the USA in 2004, 317 billion cubic feet of natural gas were consumed in the industrial production of ammonia, less than 1.5% of total U.S. annual consumption of natural gas. A 2002 report suggested that the production of ammonia consumes about 5% of global natural gas consumption, which is somewhat under 2% of world energy production.

Ammonia is produced from natural gas and air. The cost of natural gas makes up about 90% of the cost of producing ammonia. The increase in price of natural gases over the past decade, along with other factors such as increasing demand, have contributed to an increase in fertilizer price.

Contribution to Climate Change

The greenhouse gases carbon dioxide, methane and nitrous oxide are produced during the manufacture of nitrogen fertilizer. The effects can be combined into an equivalent amount of carbon dioxide. The amount varies according to the efficiency of the process. The figure for the United Kingdom is over 2 kilogrammes of carbon dioxide equivalent for each kilogramme of ammonium nitrate. Nitrogen fertilizer can be converted by soil bacteria to nitrous oxide, a greenhouse gas.

Atmosphere

Through the increasing use of nitrogen fertilizer, which was used at a rate of about 110 million tons (of N) per year in 2012, adding to the already existing amount of reactive nitrogen, nitrous oxide (N_2O) has become the third most important greenhouse gas after carbon dioxide and methane. It has a global warming potential 296 times larger than an equal mass of carbon dioxide and it also contributes to stratospheric ozone depletion. By changing processes and procedures, it is possible to mitigate some, but not all, of these effects on anthropogenic climate change.

Methane emissions from crop fields (notably rice paddy fields) are increased by the application of ammonium-based fertilizers. These emissions contribute to global climate change as methane is a potent greenhouse gas.

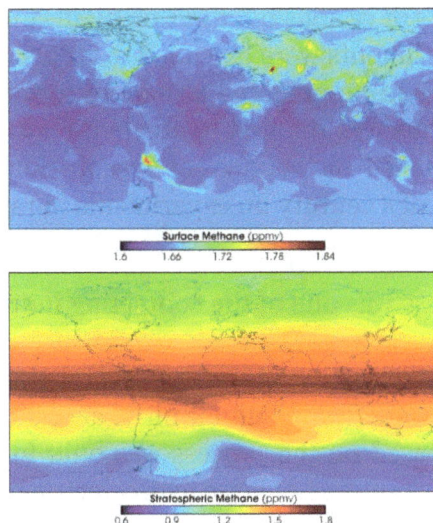

Global methane concentrations (surface and atmospheric) for 2005; note distinct plumes

Regulation

In Europe problems with high nitrate concentrations in run-off are being addressed by the European Union's Nitrates Directive. Within Britain, farmers are encouraged to manage their land more sustainably in 'catchment-sensitive farming'. In the US, high concentrations of nitrate and phosphorus in runoff and drainage water are classified as non-point source pollutants due to their diffuse origin; this pollution is regulated at state level. Oregon and Washington, both in the United States, have fertilizer registration programs with on-line databases listing chemical analyses of fertilizers.

History

Founded in 1812, Mirat, producer of manures and fertilizers, is claimed to be the oldest industrial business in Salamanca (Spain).

Management of soil fertility has been the preoccupation of farmers for thousands of years. Egyptians, Romans, Babylonians, and early Germans all are recorded as using minerals and or manure to enhance the productivity of their farms. The modern science of plant nutrition started in the 19th century and the work of German chemist Justus von Liebig, among others. John Bennet Lawes, an English entrepreneur, began to experiment on the effects of various manures on plants growing in pots in 1837, and a year or two later the experiments were extended to crops in the field. One immediate consequence was that in 1842 he patented a manure formed by treating phosphates with sulphuric acid, and thus was the first to create the artificial manure industry. In the succeeding year he enlisted the services of Joseph Henry Gilbert, with whom he carried on for more than half a century on experiments in raising crops at the Institute of Arable Crops Research.

The Birkeland–Eyde process was one of the competing industrial processes in the beginning of nitrogen based fertilizer production. This process was used to fix atmospheric nitrogen (N_2) into nitric acid (HNO_3), one of several chemical processes generally referred to as nitrogen fixation. The resultant nitric acid was then used as a source of nitrate (NO_3^-). A factory based on the process was built in Rjukan and Notodden in Norway, combined with the building of large hydroelectric power facilities.

The 1910s and 1920s witness the rise of the Haber process and the Ostwald process. The Haber process produces ammonia (NH_3) from methane (CH_4) gas and molecular nitrogen (N_2). The ammonia from the Haber process is then converted into nitric acid (HNO_3) in the Ostwald process. The development of synthetic fertilizer has significantly supported global population growth — it has been estimated that almost half the people on the Earth are currently fed as a result of synthetic nitrogen fertilizer use.

The use of commercial fertilizers has increased steadily in the last 50 years, rising almost 20-fold to the current rate of 100 million tonnes of nitrogen per year. Without commercial fertilizers it is estimated that about one-third of the food produced now could not be produced. The use of phosphate fertilizers has also increased from 9 million tonnes per year in 1960 to 40 million tonnes per year in 2000. A maize crop yielding 6–9 tonnes of grain per hectare (2.5 acres) requires 31–50 kilograms (68–110 lb) of phosphate fertilizer to be applied; soybean crops require about half, as 20–25 kg per hectare. Yara International is the world's largest producer of nitrogen-based fertilizers.

Controlled-nitrogen-release technologies based on polymers derived from combining urea and formaldehyde were first produced in 1936 and commercialized in 1955. The early product had 60 percent of the total nitrogen cold-water-insoluble, and the unreacted (quick-release) less than 15%. Methylene ureas were commercialized in the 1960s and 1970s, having 25% and 60% of the nitrogen as cold-water-insoluble, and unreacted urea nitrogen in the range of 15% to 30%.

In the 1960s, the Tennessee Valley Authority National Fertilizer Development Center began developing sulfur-coated urea; sulfur was used as the principal coating material because of its low cost and its value as a secondary nutrient. Usually there is another wax or polymer which seals the sulfur; the slow-release properties depend on the degradation of the secondary sealant by soil microbes as well as mechanical imperfections (cracks, etc.) in the sulfur. They typically provide 6 to 16 weeks of delayed release in turf applications. When a hard polymer is used as the secondary coating, the properties are a cross between diffusion-controlled particles and traditional sulfur-coated.

Chemicals Compounds Used in Fertilizer

Ammonium Nitrate

Ammonium nitrate is a chemical compound, the nitrate salt of the ammonium cation. It has the chemical formula NH_4NO_3, simplified to $N_2H_4O_3$. It is a white crystalline solid and is highly soluble in water. It is predominantly used in agriculture as a high-nitrogen fertilizer. Its other major use is as a component of explosive mixtures used in mining, quarrying, and civil construction. It is the major constituent of ANFO, a popular industrial explosive which accounts for 80% of explosives used in North America; similar formulations have been used in improvised explosive devices. Many countries are phasing out its use in consumer applications due to concerns over its potential for misuse.

Occurrence

Ammonium nitrate is found as a natural mineral (ammonia nitre—the ammonium analogue of saltpetre and other nitre minerals such as sodium nitrate) in the driest regions of the Atacama Desert in Chile, often as a crust on the ground and/or in conjunction with other nitrate, chlorate, iodate, and halide minerals. Ammonium nitrate was mined there in the past, but virtually 100% of the chemical now used is synthetic.

Production

The industrial production of ammonium nitrate entails the acid-base reaction of ammonia with nitric acid:

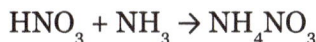

$$HNO_3 + NH_3 \rightarrow NH_4NO_3$$

Ammonia is used in its anhydrous form (i.e., gas form) and the nitric acid is concentrated. This reaction is violent owing to its highly exothermic nature. After the solution is formed, typically at about 83% concentration, the excess water is evaporated to an ammonium nitrate (AN) content of 95% to 99.9% concentration (AN melt), depending on grade. The AN melt is then made into "prills" or small beads in a spray tower, or into granules by spraying and tumbling in a rotating drum. The prills or granules may be further dried, cooled, and then coated to prevent caking. These prills or granules are the typical AN products in commerce.

The ammonia required for this process is obtained by the Haber process from nitrogen and hydrogen. Ammonia produced by the Haber process is oxidized to nitric acid by the Ostwald process. Another production method is a variant of the Odda process:

$$Ca(NO_3)_2 + 2\ NH_3 + CO_2 + H_2O \rightarrow 2\ NH_4NO_3 + CaCO_3$$

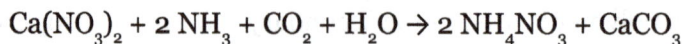

The products, calcium carbonate and ammonium nitrate, may be separately purified or sold combined as calcium ammonium nitrate.

Ammonium nitrate can also be made via metathesis reactions:

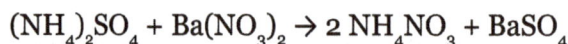

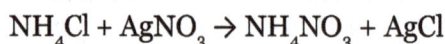

$$(NH_4)_2SO_4 + Ba(NO_3)_2 \rightarrow 2\ NH_4NO_3 + BaSO_4$$

$$NH_4Cl + AgNO_3 \rightarrow NH_4NO_3 + AgCl$$

Properties

Reactions

Ammonium nitrate reacts with metal hydroxides, releasing ammonia and forming alkali metal nitrate:

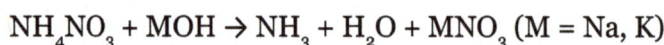

$$NH_4NO_3 + MOH \rightarrow NH_3 + H_2O + MNO_3\ (M = Na, K)$$

Ammonium nitrate leaves no residue when heated:

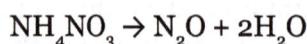

$$NH_4NO_3 \rightarrow N_2O + 2H_2O$$

Ammonium nitrate is also formed in the atmosphere from emissions of NO, SO_2, and NH_3, and is a secondary component of PM10.

Crystalline Phases

Transformations of the crystal states due to changing conditions (temperature, pressure) affect the physical properties of ammonium nitrate. These crystalline states have been identified:

System	Temperature (°C)	State	Volume change (%)
	> 169.6	liquid	
I	169.6 to 125.2	cubic	+2.1
II	125.2 to 84.2	tetragonal	−1.3
III	84.2 to 32.3	α-rhombic	+3.6
IV	32.3 to −16.8	β-rhombic	−2.9
V	−16.8	tetragonal	

The type V crystal is a quasicubic form related to caesium chloride, the nitrogen atoms of the nitrate anions and the ammonium cations are at the sites in a cubic array where Cs and Cl would be in the CsCl lattice.

Applications

Fertilizer

Ammonium nitrate is an important fertilizer with the NPK rating 34-0-0 (34% nitrogen). It is less concentrated than urea (46-0-0), giving ammonium nitrate a slight transportation disadvantage. Ammonium nitrate's advantage over urea is that it is more stable and does not rapidly lose nitrogen to the atmosphere. During warm weather it is best to apply urea soon before rain is expected or to cover it with soil to minimize nitrogen loss.

Explosives

Ammonium nitrate is not, on its own, an explosive, but it readily forms explosive mixtures with varying properties when combined with primary explosives such as azides or with fuels such aluminum powder or fuel oil.

Mixture With Fuel Oil

ANFO is a mixture of 94% ammonium nitrate ("AN") and 6% fuel oil ("FO") widely used as a bulk industrial explosive. It is used in coal mining, quarrying, metal mining, and civil construction in undemanding applications where the advantages of ANFO's low cost and ease of use matter more than the benefits offered by conventional industrial explosives, such as water resistance, oxygen balance, high detonation velocity, and performance in small diameters.

Terrorism

Ammonium nitrate-based explosives were used in the Oklahoma City Bombing in 1995 and 2011 Delhi bombings, the 2013 Hyderabad blasts, and the 2011 bombing in Oslo.

In November 2009, a ban on ammonium sulfate, ammonium nitrate, and calcium ammonium nitrate fertilizers was imposed in the former Malakand Division—comprising the Upper Dir, Lower Dir, Swat, Chitral, and Malakand districts of the North West Frontier Province (NWFP) of Pakistan—by the NWFP government, following reports that those chemicals were used by militants to make explosives. Due to these bans, "Potassium chlorate — the stuff that makes matches catch fire — has surpassed fertilizer as the explosive of choice for insurgents."

Niche Uses

Ammonium nitrate is used in some instant cold packs, as its dissolution in water is highly endothermic. It also was used, in combination with independently explosive "fuels" such as guanidine nitrate, as a cheaper (but less stable) alternative to 5-aminotetrazole in the inflaters of airbags manufactured by Takata Corporation, which were recalled as unsafe after killing 14 people.

Safety, Handling, and Storage

Health and safety data are shown on the safety data sheets available from suppliers and found on the internet. In response to several explosions resulting in the deaths of numerous people, U.S. agencies of Environmental Protection (EPA), Occupational Health and Safety (OSHA) and the Bureau of Alcohol, Tobacco and Firearms jointly issued safety guidelines.

Heating or any ignition source may cause violent combustion or explosion. Ammonium nitrate reacts with combustible and reducing materials as it is a strong oxidant. Although it is mainly used for fertilizer, it can be used for explosives. It was sometimes used to blast away earth to make farm ponds. Ammonium nitrate is also used to modify the detonation rate of other explosives, such as trinitrotoluene in the form of amatol.

Numerous safety guidelines are available for storing and handling ammonium nitrate. It should not be stored near combustible substances. Ammonium nitrate is incompatible with certain substances such as chlorates, mineral acids and metal sulfides, contact with which can lead to vigorous or even violent decomposition.

Ammonium nitrate has a critical relative humidity of 59.4%, above which it will absorb moisture from the atmosphere. Therefore, it is important to store ammonium nitrate in a tightly sealed container. Otherwise, it can coalesce into a large, solid mass. Ammonium nitrate can absorb enough moisture to liquefy. Blending ammonium nitrate with certain other fertilizers can lower the critical relative humidity.

The potential for use of the material as an explosive has prompted regulatory measures. For example, in Australia, the Dangerous Goods Regulations came into effect in August 2005 to enforce licensing in dealing with such substances. Licenses are granted only to applicants (industry) with appropriate security measures in place to prevent any misuse. Additional uses such as education and research purposes may also be considered, but individual use will not. Employees of those with licenses to deal with the substance are still required to be supervised by authorized personnel and are required to pass a security and national police check before a license may be granted.

Health Hazards

Health and safety data are shown on the material safety data sheets, which are available from suppliers and can be found on the internet.

Ammonium nitrate is not very hazardous to health and is usually used in fertilizer products.

Ammonium nitrate has an LD_{50} of 2217 mg/kg, which for comparison is about two-thirds that of table salt.

Acute Health Effects

Short-term exposure to ammonium nitrate can cause symptoms ranging from minor irritation to nausea, vomiting, gastric irritation, headaches, dizziness, and hypertension.

Area of exposure	Hazard level
Ingestion	Moderately hazardous
Skin contact	Moderately hazardous (irritant)
Eye contact	Moderately hazardous
Inhalation	Moderately hazardous

Disasters

Ammonium nitrate decomposes into the gases nitrous oxide and water vapor when heated (not an explosive reaction); however, it can be induced to decompose explosively by detonation. Large stockpiles of the material can be a major fire risk due to their supporting oxidation, and may also detonate, as happened in the Texas City disaster of 1947, which led to major changes in the regulations for storage and handling.

Two major classes of incidents resulting in explosions are:

- The explosion happens by the mechanism of shock-to-detonation transition. The initiation happens by an explosive charge going off in the mass, by the detonation of a shell thrown into the mass, or by detonation of an explosive mixture in contact with the mass. The examples are Kriewald, Morgan (present-day Sayreville, New Jersey), Oppau, and Tessenderlo.

- The explosion results from a fire that spreads into the ammonium nitrate itself (Texas City, Brest, Oakdale PA), or from a mixture of ammonium nitrate with a combustible material during the fire (Repauno, Cherokee, Nadadores). The fire must be confined at least to a degree for successful transition from a fire to an explosion (a phenomenon known as "deflagration-to-detonation transition"). Pure, compact AN is stable and very difficult to ignite, and numerous cases exist when even impure AN did not explode in a fire.

Ammonium nitrate was suspected as the explosive responsible for the fertilizer plant explosion in West, Texas on April 17, 2013. Investigators said they believe it exploded following a fire that began in the plant's office.

Calcium Nitrate

Calcium nitrate, also called *Norgessalpeter* (Norwegian saltpeter), is the inorganic compound with the formula $Ca(NO_3)_2$. This colourless salt absorbs moisture from the air and is commonly found as a tetrahydrate. It is mainly used as a component in fertilizers but has other applications. Nitrocalcite is the name for a mineral which is a hydrated calcium nitrate that forms as an efflorescence where manure contacts concrete or limestone in a dry environment as in stables or caverns. A variety of related salts are known including calcium ammonium nitrate decahydrate and calcium potassium nitrate decahydrate.

Production and eactivity

Norgessalpeter was the first synthetic nitrogen fertilizer compound to be manufactured. Production began at Notodden, Norway in 1905 by the Birkeland–Eyde process. Most of the world's calcium nitrate is now made in Porsgrunn.

It is produced by treating limestone with nitric acid, followed by neutralization with ammonia:

$$CaCO_3 + 2\ HNO_3 \rightarrow Ca(NO_3)_2 + CO_2 + H_2O$$

It is also an intermediate product of the Odda Process:

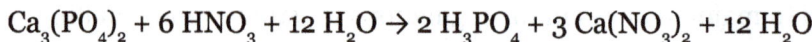

$$Ca_3(PO_4)_2 + 6\ HNO_3 + 12\ H_2O \rightarrow 2\ H_3PO_4 + 3\ Ca(NO_3)_2 + 12\ H_2O$$

It can also be prepared from an aqueous solution of ammonium nitrate, and calcium hydroxide:

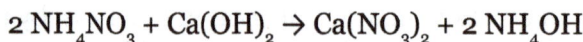

$$2\ NH_4NO_3 + Ca(OH)_2 \rightarrow Ca(NO_3)_2 + 2\ NH_4OH$$

Like related alkaline earth metal nitrates, calcium nitrate decomposes upon heating (starting at 500 °C) to release nitrogen dioxide:

$$2\ Ca(NO_3)_2 \rightarrow 2\ CaO + 4\ NO_2 + O_2\ \Delta H = 369\ kJ/mol$$

Applications

Use in Agriculture

As of 1978, only 170,000 tons/year were produced for applications in fertilizers. The fertilizer grade (15.5-0-0 + 19% Ca) is popular in the greenhouse and hydroponics trades; it contains ammonium nitrate and water, as the "double salt" $5Ca(NO_3)_2$. $NH_4NO_3 \cdot 10H_2O$. This is called calcium ammonium nitrate. Formulations lacking ammonia are also known: $Ca(NO_3)_2 \cdot 4H_2O$ (11.9-0-0 + 16.9%Ca). A liquid formulation (9-0-0 + 11% Ca) is also offered. An anhydrous, air-stable derivative is the urea complex $Ca(NO_3)_2 \cdot 4[OC(NH_2)_2]$, which has been sold as Cal-Urea.

Calcium nitrate is also used to control certain plant diseases. For example, dilute calcium nitrate (and calcium chloride) sprays are used to control bitter pit and cork spot in apple trees.

Waste Water Treatment

Calcium nitrate is used in waste water pre-conditioning for odour emission prevention. The waste water pre-conditioning is based on establishing an anoxic biology in the waste water system. In the presence of nitrate, the metabolism for sulfates stops, thus preventing formation of hydrogen sulphide. Additionally easy degradable organic matter is consumed, which otherwise can cause anaerobic conditions downstream as well as odour emissions itself. The concept is also applicable for surplus sludge treatment.

Concrete

Calcium nitrate is used in set accelerating concrete admixtures. This use with concrete and mortar is based on two effects. The calcium ion accelerates formation of calcium hydroxide and thus precipitation and setting. This effect is used also in cold weather concreting agents as well as some combined plasticizers. The nitrate ion leads to formation of iron hydroxide, whose protective layer reduces corrosion of the concrete reinforcement.

Latex Coagulant

Calcium nitrate is a very common coagulant in latex production, especially in dipping processes. Dissolved calcium nitrate is a part of the dipping bath solution. The warm former is dipped into the coagulation liquid and a thin film of the dipping liquid remains on the former. When now dipping the former into the latex the calcium nitrate will break up the stabilization of the latex solution and the latex will coagulate on the former.

Cold Packs

The dissolution of calcium nitrate tetrahydrate is highly endothermic (cooling). For this reason, calcium nitrate tetrahydrate is sometimes used for regenerable cold packs.

Molten salts for Heat Transfer and Storage

Calcium nitrate can be used as a part of molten salt mixtures. Typical are binary mixtures of calcium nitrate and potassium nitrate or ternary mixtures including also sodium nitrate. Those molten salts can be used to replace thermo oil in concentrated solar power plants for the heat transfer, but mostly those are used in heat storage.

Potassium Nitrate

Potassium nitrate is a chemical compound with the chemical formula KNO_3. It is an ionic salt of potassium ions K^+ and nitrate ions NO_3^-, and is therefore an alkali metal nitrate.

It occurs as a mineral niter and is a natural solid source of nitrogen. Potassium nitrate is one of several nitrogen-containing compounds collectively referred to as saltpeter or saltpetre.

Major uses of potassium nitrate are in fertilizers, tree stump removal, rocket propellants and fireworks. It is one of the major constituents of gunpowder (blackpowder) and has been used since the Middle Ages as a food preservative.

Properties

Potassium nitrate has an orthorhombic crystal structure at room temperature, which transforms to a trigonal system at 129 °C (264 °F).

Potassium nitrate is moderately soluble in water, but its solubility increases with temperature. The aqueous solution is almost neutral, exhibiting pH 6.2 at 14 °C (57 °F) for a 10% solution of commercial powder. It is not very hygroscopic, absorbing about 0.03% water in 80% relative humidity over 50 days. It is insoluble in alcohol and is not poisonous; it can react explosively with reducing agents, but it is not explosive on its own.

Thermal Decomposition

Between 550–790 °C (1,022–1,454 °F), potassium nitrate reaches a temperature dependent equilibrium with potassium nitrite:

$$2 \; KNO_3 \rightleftharpoons 2 \; KNO_2 + O_2$$

History of Production

From Mineral Sources

The earliest known complete purification process for potassium nitrate was outlined in 1270 by the chemist and engineer Hasan al-Rammah of Syria in his book *al-Furusiyya wa al-Manasib al-Harbiyya* (*The Book of Military Horsemanship and Ingenious War Devices*). In this book, al-Rammah describes first the purification of *barud* (crude saltpeter mineral) by boiling it with minimal water and using only the hot solution, then the use of potassium carbonate (in the form of wood ashes) to remove calcium and magnesium by precipitation of their carbonates from this solution, leaving a solution of purified potassium nitrate, which could then be dried. This was used for the manufacture of gunpowder and explosive devices. The terminology used by al-Rammah indicated a Chinese origin for the gunpowder weapons about which he wrote.

At least as far back as 1845, Chilean saltpeter deposits were exploited in Chile and California, USA.

From Caves

A major natural source of potassium nitrate was the deposits crystallizing from cave walls and the accumulations of bat guano in caves. Extraction is accomplished by immersing the guano in water for a day, filtering, and harvesting the crystals in the filtered water. Traditionally, guano was the source used in Laos for the manufacture of gunpowder for *Bang Fai* rockets.

LeConte

Perhaps the most exhaustive discussion of the production of this material is the 1862 LeConte text. He was writing with the express purpose of increasing production in the Confederate States to support their needs during the American Civil War. Since he was calling for the assistance of rural farming communities, the descriptions and instructions are both simple and explicit. He details the "French Method", along with several variations, as well as a "Swiss method". N.B. Many references have been made to a method using only straw and urine, but there is no such method in this work.

French Method

Niter-beds are prepared by mixing manure with either mortar or wood ashes, common earth and organic materials such as straw to give porosity to a compost pile typically 4 feet (1.2 m) high, 6 feet (1.8 m) wide, and 15 feet (4.6 m) long. The heap was usually under a cover from the rain, kept moist with urine, turned often to accelerate the decomposition, then finally leached with water after approximately one year, to remove the soluble calcium nitrate which was then converted to potassium nitrate by filtering through the potash.

Swiss Method

LeConte describes a process using only urine and not dung, referring to it as the *Swiss method*. Urine is collected directly, in a sandpit under a stable. The sand itself is dug out and leached for nitrates which were then converted to potassium nitrate via potash, as above.

From Nitric Acid

From 1903 until the World War I era, potassium nitrate for black powder and fertilizer was produced on an industrial scale from nitric acid produced via the Birkeland–Eyde process, which used an electric arc to oxidize nitrogen from the air. During World War I the newly industrialized Haber process (1913) was combined with the Ostwald process after 1915, allowing Germany to produce nitric acid for the war after being cut off from its supplies of mineral sodium nitrates from Chile

Production

Potassium nitrate can be made by combining ammonium nitrate and potassium hydroxide.

$$NH_4NO_3 \text{ (aq)} + KOH \text{ (aq)} \rightarrow NH_3 \text{ (g)} + KNO_3 \text{ (aq)} + H_2O \text{ (l)}$$

An alternative way of producing potassium nitrate without a by-product of ammonia is to combine ammonium nitrate and potassium chloride, easily obtained as a sodium-free salt substitute.

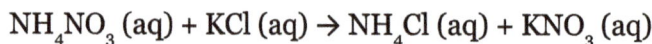

$$NH_4NO_3 \text{ (aq)} + KCl \text{ (aq)} \rightarrow NH_4Cl \text{ (aq)} + KNO_3 \text{ (aq)}$$

Potassium nitrate can also be produced by neutralizing nitric acid with potassium hydroxide. This reaction is highly exothermic.

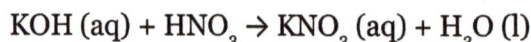

$$KOH \text{ (aq)} + HNO_3 \rightarrow KNO_3 \text{ (aq)} + H_2O \text{ (l)}$$

On industrial scale it is prepared by the double displacement reaction between sodium nitrate and potassium chloride.

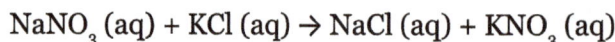

$$NaNO_3 \text{ (aq)} + KCl \text{ (aq)} \rightarrow NaCl \text{ (aq)} + KNO_3 \text{ (aq)}$$

Uses

Potassium nitrate has a wide variety of uses, largely as a source of nitrate.

Nitric Acid Production

Historically, nitric acid was produced by combining sulfuric acid with nitrates such as saltpeter. In modern times this is reversed: nitrates are produced from nitric acid produced via the Ostwald process.

Oxidizer

The most famous use of potassium nitrate is probably as the oxidizer in blackpowder. From the most ancient times through the late 1880s, blackpowder provided the explosive power for all the world's firearms. After that time, small arms and large artillery increasingly began to depend on cordite, a smokeless powder. Blackpowder remains in use today in black powder rocket motors, but also in combination with other fuels like sugars in "rocket candy". It is also used in fireworks such as smoke bombs. It is also added to cigarettes to maintain an even burn of the tobacco and is used to ensure complete combustion of paper cartridges for cap and ball revolvers. It can also be heated to several hundred degrees to be used for niter bluing, which is less durable than other forms of protective oxidation, but allows for specific and often beautiful coloration of steel parts, such as screws, pins, and other small parts of firearms.

Food Preservation

In the process of food preservation, potassium nitrate has been a common ingredient of salted meat since the Middle Ages, but its use has been mostly discontinued because of inconsistent results compared to more modern nitrate and nitrite compounds. Even so, saltpeter is still used in some food applications, such as *charcuterie* and the brine used to make corned beef. When used as a food additive in the European Union, the compound is referred to as E252; it is also approved for use as a food additive in the USA and Australia and New Zealand (where it is listed under its INS number 252). Although nitrate salts have been suspected of producing the carcinogen nitrosamine, both sodium and potassium nitrates and nitrites have been added to meats in the US since 1925, and nitrates and nitrites have not been removed from preserved meat products because nitrite and nitrate inhibits the germination of C. botulinum endospores, and thus prevents botulism from bacterial toxin that may otherwise be produced in certain preserved meat products.

Food Preparation

In West African cuisine, potassium nitrate (salt petre) is widely used as a thickening agent in soups and stews such as Okra soup and Isi ewu. It is also used to soften food and reduce cooking time when boiling beans and tough meat. Salt petre is also an essential ingredient in making special porridges such as *kunun kanwa* literally translated from the Hausa language as 'salt petre porridge'. In the Shetland Islands (UK) it is used in the curing of mutton to make "reestit" mutton, a local delicacy.

Fertilizer

Potassium nitrate is used in fertilizers as a source of nitrogen and potassium – two of the macronutrients for plants. When used by itself, it has an NPK rating of 13-0-44.

Pharmacology

- Used in some toothpastes for sensitive teeth. Recently, the use of potassium nitrate in toothpastes for treating sensitive teeth has increased and it may be an effective treatment.

- Used historically to treat asthma. Used in some toothpastes to relieve asthma symptoms.

- Used in Thailand as main ingredient in kidney tablets to relieve the symptoms of cystitis, pyelitis and urethritis.

- Combats high blood pressure and was once used as a hypotensive.

Other uses

- Electrolyte in a salt bridge

- Active ingredient of condensed aerosol fire suppression systems. When burned with the free radicals of a fire's flame, it produces potassium carbonate.

- Works as an aluminium cleaner.

- Component (usually about 98%) of some tree stump removal products. It accelerates the natural decomposition of the stump by supplying nitrogen for the fungi attacking the wood of the stump.

- In heat treatment of metals as a medium temperature molten salt bath, usually in combination with sodium nitrite. A similar bath is used to produce a durable blue/black finish typically seen on firearms. Its oxidizing quality, water solubility, and low cost make it an ideal short-term rust inhibitor.

- To induce flowering of mango trees in the Philippines.

- Thermal storage medium in power generation systems. Sodium and potassium nitrate salts are stored in a molten state with the solar energy collected by the heliostats at the Gemasolar Thermosolar Plant. Ternary salts, with the addition of calcium nitrate or lithium nitrate, have been found to improve the heat storage capacity in the molten salts.

In Folklore and Popular Culture

Potassium nitrate was once thought to induce impotence, and is still falsely rumored to be in institutional food (such as military fare) as an anaphrodisiac; however, there is no scientific evidence for such properties.

Monocalcium Phosphate

Monocalcium phosphate is an inorganic compound with the chemical formula $Ca(H_2PO_4)_2$ ("ACMP" or "CMP-A" for anhydrous monocalcium phosphate). It is commonly found as the monohydrate (""MCP" or "MCP-M"), $Ca(H_2PO_4)_2 \cdot H_2O$ (CAS# 10031-30-8). Both salts are colourless solids. They are used mainly as superphosphate fertilizers and are also popular leavening agents.

Preparation

Material of relatively high purity, as required for baking, is produced by treating calcium hydroxide with phosphoric acid:

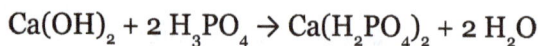

$$Ca(OH)_2 + 2 H_3PO_4 \rightarrow Ca(H_2PO_4)_2 + 2 H_2O$$

Samples of $Ca(H_2PO_4)_2$ tends to convert to dicalcium phosphate:

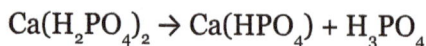

$$Ca(H_2PO_4)_2 \rightarrow Ca(HPO_4) + H_3PO_4$$

Applications

Use in Fertilizers

Superphosphate fertilizers are produced by treatment of "phosphate rock" with acids. Using phosphoric acid, fluorapatite is converted to $Ca(H_2PO_4)_2$:

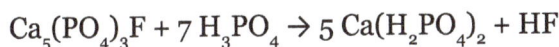

$$Ca_5(PO_4)_3F + 7 H_3PO_4 \rightarrow 5 Ca(H_2PO_4)_2 + HF$$

This solid is called triple superphosphate. Several million tons are produced annually for use as fertilizers. Residual HF typically reacts with silicate minerals co-mingled with the phosphate ores to produce hydrofluorosilicic acid (H_2SiF_6). The majority of the hexafluorosilicic acid is converted to aluminium fluoride and cryolite for the processing of aluminium. These materials are central to the conversion of aluminium ore into aluminium metal.

When sulfuric acid is used, the product contains phosphogypsum ($CaSO_4 \cdot 2H_2O$) and is called single superphosphate.

Use as Leavening Agent

Calcium dihydrogen phosphate is used in the food industry as a leavening agent, i.e., to cause baked goods to rise. Because it is acidic, when combined with an alkali ingredient, commonly sodium bicarbonate (baking soda) or potassium bicarbonate, it reacts to produce carbon dioxide and a salt. Outward pressure of the carbon dioxide gas causes the rising effect. When combined in a ready-made baking powder, the acid and alkali ingredients are included in the right proportions such that they will exactly neutralize each other and not significantly affect the overall pH of the product. AMCP and MCP are fast acting, releasing most carbon dioxide within minutes of mixing. It is popularly used in pancake mixes. In double acting baking powders, MCP is often combined with the slow acting acid sodium acid pyrophosphate (SAPP).

Organic Fertilizer

A cement reservoir containing cow manure mixed with water. This is common in rural Hainan Province, China. Note the bucket on a stick that the farmer uses to apply the mixture.

Organic fertilizers are fertilizers derived from animal matter, human excreta or vegetable matter. (e.g. compost, manure). Naturally occurring organic fertilizers include animal wastes from meat processing, peat, manure, slurry, and guano.

In contrast, the majority of fertilizers used in commercial farming are extracted from minerals (e.g., phosphate rock) or produced industrially (e.g., ammonia).

Compost bin for small-scale production of organic fertilizer

A large commercial compost operation

Examples and Sources

The main organic fertilizers are, peat, animal wastes (often from slaughter houses), plant wastes from agriculture, and treated sewage sludge.

Mineral

The main source of organic fertilizer is peat, an immature precursor to coal. Peat itself offers no nutritional value to the plants, but improves the soil by aeration and absorbing water.

Peat is the most widely used organic fertilizer.

Mined powdered limestone, rock phosphate, and Chilean saltpeter are inorganic (not of biologic origins) compounds, which can be energetically intensive to harvest.

Animal Sources

These materials include the products of the slaughter of animals. Bloodmeal, bone meal, hides, hoofs, and horns are typical precursors. fish meal, and feather meal are other sources.

Chicken litter, which consists of chicken manure mixed with sawdust, is an organic fertilizer that has been shown to better condition soil for harvest than synthesized fertilizer. Researchers at the Agricultural Research Service (ARS) studied the effects of using chicken litter, an organic fertilizer, versus synthetic fertilizers on cotton fields, and found that fields fertilized with chicken litter had a 12% increase in cotton yields over fields fertilized with synthetic fertilizer. In addition to higher yields, researchers valued commercially sold chicken litter at a $17/ton premium (to a total valuation of $78/ton) over the traditional valuations of $61/ton due to value added as a soil conditioner.

Plant

Processed organic fertilizers include compost, humic acid, amino acids, and seaweed extracts. Other examples are natural enzyme-digested proteins. Decomposing crop residue (green manure) from prior years is another source of fertility.

Other ARS studies have found that algae used to capture nitrogen and phosphorus runoff from agricultural fields can not only prevent water contamination of these nutrients, but also can be used as an organic fertilizer. ARS scientists originally developed the "algal turf scrubber" to reduce nutrient runoff and increase quality of water flowing into streams, rivers, and lakes. They found that this nutrient-rich algae, once dried, can be applied to cucumber and corn seedlings and result in growth comparable to that seen using synthetic fertilizers.

Treated Sewage Sludge

Although night soil (from human excreta) was a traditional organic fertilizer, the main source of this type is nowadays treated sewage sludge, also known as biosolids.

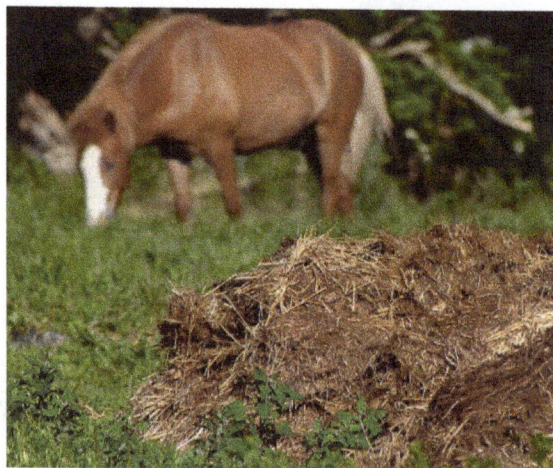

Decomposing animal manure, an organic fertilizer source

Biosolids as soil amendment is only available to less than 1% of US agricultural land. Industrial pollutants in sewage sludge prevents recycling it as fertilizer. The USDA prohibits use of sewage sludge in organic agricultural operations in the U.S. due to industrial pollution, pharmaceuticals, hormones, heavy metals, and other factors. The USDA now requires 3rd-party certification of high-nitrogen liquid organic fertilizers sold in the U.S.

Sewage sludge use in organic agricultural operations in the U.S. has been extremely limited and rare due to USDA prohibition of the practice (due to toxic metal accumulation, among other factors).

Urine

Animal sourced urea and urea-formaldehyde from urine are suitable for organic agriculture; however, synthetically produced urea is not. The common thread that can be seen through these examples is that *organic* agriculture attempts to define itself through minimal processing (e.g., via chemical energy such as petroleum , as well as being naturally occurring or via natural biological processes such as composting.

Others

- Alfalfa
- Ash
- Blood meal

- Bone meal

- Compost

- Cover crops

- Fish emulsion

- Fish meal

- Manure

- Rock phosphate

- Raw Langbeinite

- Rockdust

- Unprocessed natural potassium sulfate

- Wood chips/sawdust

- PROM

Urea

Urea, also known as carbamide, is an organic compound with the chemical formula $CO(NH_2)_2$. The molecule has two $-NH_2$ groups joined by a carbonyl (C=O) functional group.

Urea serves an important role in the metabolism of nitrogen-containing compounds by animals, and is the main nitrogen-containing substance in the urine of mammals. It is a colorless, odorless solid, highly soluble in water, and practically non-toxic (LD_{50} is 15 g/kg for rats). Dissolved in water, it is neither acidic nor alkaline. The body uses it in many processes, most notably nitrogen excretion. The liver forms it by combining two ammonia molecules (NH_3) with a carbon dioxide (CO_2) molecule in the urea cycle. Urea is widely used in fertilizers as a source of nitrogen and is an important raw material for the chemical industry.

Friedrich Wöhler's discovery in 1828 that urea can be produced from inorganic starting materials was an important conceptual milestone in chemistry. It showed for the first time that a substance previously known only as a byproduct of life could be synthesized in the laboratory without biological starting materials, contradicting the widely held doctrine of vitalism.

It is on the WHO Model List of Essential Medicines, the most important medications needed in a basic health system.

Uses

Agriculture

A process plant in Bangladesh, that commercially produces Urea as fertilizer by using Methane as main raw material

More than 90% of world industrial production of urea is destined for use as a nitrogen-release fertilizer. Urea has the highest nitrogen content of all solid nitrogenous fertilizers in common use. Therefore, it has the lowest transportation costs per unit of nitrogen nutrient. The standard crop-nutrient rating (NPK rating) of urea is 46-0-0.

Many soil bacteria possess the enzyme urease, which catalyzes conversion of urea to ammonia or ammonium ion and bicarbonate ion. Thus urea fertilizers rapidly transform to the ammonium form in soils. Among the soil bacteria known to carry urease, some ammonia-oxidizing bacteria (AOB), such as species of *Nitrosomonas*, can also assimilate the carbon dioxide the reaction releases to make biomass via the Calvin Cycle, and harvest energy by oxidizing ammonia (the other product of urease) to nitrite, a process termed nitrification. Nitrite-oxidizing bacteria, especially *Nitrobacter*, oxidize nitrite to nitrate, which is extremely mobile in soils because of its negative charge and is a major cause of water pollution from agriculture. Ammonium and nitrate are readily absorbed by plants, and are the dominant sources of nitrogen for plant growth. Urea is also used in many multi-component solid fertilizer formulations. Urea is highly soluble in water and is therefore also very suitable for use in fertilizer solutions (in combination with ammonium nitrate: UAN), e.g., in 'foliar feed' fertilizers. For fertilizer use, granules are preferred over prills because of their narrower particle size distribution, which is an advantage for mechanical application.

The most common impurity of synthetic urea is biuret, which impairs plant growth.

Urea is usually spread at rates of between 40 and 300 kg/ha but rates vary. Smaller applications incur lower losses due to leaching. During summer, urea is often spread just before or during rain to minimize losses from volatilization (a process wherein nitrogen is lost to the atmosphere as ammonia gas).

Because of the high nitrogen concentration in urea, it is very important to achieve an even spread. The application equipment must be correctly calibrated and properly used. Drilling must not occur on contact with or close to seed, due to the risk of germination damage. Urea dissolves in water for application as a spray or through irrigation systems.

In grain and cotton crops, urea is often applied at the time of the last cultivation before planting. In high rainfall areas and on sandy soils (where nitrogen can be lost through leaching) and where good in-season rainfall is expected, urea can be side- or top-dressed during the growing season. Top-dressing is also popular on pasture and forage crops. In cultivating sugarcane, urea is side-dressed after planting, and applied to each ratoon crop.

In irrigated crops, urea can be applied dry to the soil, or dissolved and applied through the irrigation water. Urea dissolves in its own weight in water, but becomes increasingly difficult to dissolve as the concentration increases. Dissolving urea in water is endothermic—the solution temperature falls when urea dissolves.

As a practical guide, when preparing urea solutions for fertigation (injection into irrigation lines), dissolve no more than 3 g urea per 1 L water.

In foliar sprays, urea concentrations of between 0.5% and 2.0% are often used in horticultural crops. Low-biuret grades of urea are often indicated.

Urea absorbs moisture from the atmosphere and therefore is typically stored either in closed or sealed bags on pallets or, if stored in bulk, under cover with a tarpaulin. As with most solid fertilizers, storage in a cool, dry, well-ventilated area is recommended.

Overdose or placing urea near seed is harmful.

Chemical Industry

Urea is a raw material for the manufacture of two main classes of materials: urea-formaldehyde resins and urea-melamine-formaldehyde used in marine plywood.

Explosives

Urea can be used to make urea nitrate, a high explosive that is used industrially and as part of some improvised explosive devices. It is a stabilizer in nitrocellulose explosives.

Automobile Systems

Urea is used in SNCR and SCR reactions to reduce the NO_x pollutants in exhaust gases from combustion from Diesel, dual fuel, and lean-burn natural gas engines. The BlueTec system, for example, injects a water-based urea solution into the exhaust system. The ammonia produced by the hydrolysis of the urea reacts with the nitrogen oxide emissions and is converted into nitrogen and water within the catalytic converter. Trucks and cars using these catalytic converters need to carry a supply of diesel exhaust fluid (DEF, also known as AdBlue), a mixture of urea and water.

Laboratory Uses

Urea in concentrations up to 10 M is a powerful protein denaturant as it disrupts the noncovalent bonds in the proteins. This property can be exploited to increase the solubility of some proteins. A mixture of urea and choline chloride is used as a deep eutectic solvent, a type of ionic liquid.

Urea can in principle serve as a hydrogen source for subsequent power generation in fuel cells. Urea present in urine/wastewater can be used directly (though bacteria normally quickly degrade urea.) Producing hydrogen by electrolysis of urea solution occurs at a lower voltage (0.37 V) and thus consumes less energy than the electrolysis of water (1.2V).

Urea in concentrations up to 8 M can be used to make fixed brain tissue transparent to visible light while still preserving fluorescent signals from labeled cells. This allows for much deeper imaging of neuronal processes than previously obtainable using conventional one photon or two photon confocal microscopes.

Medical Use

Urea-containing creams are used as topical dermatological products to promote rehydration of the skin. Urea 40% is indicated for psoriasis, xerosis, onychomycosis, ichthyosis, eczema, keratosis, keratoderma, corns, and calluses. If covered by an occlusive dressing, 40% urea preparations may also be used for nonsurgical debridement of nails. Urea 40% "dissolves the intercellular matrix" of the nail plate. Only diseased or dystrophic nails are removed, as there is no effect on healthy portions of the nail. This drug is also used as an earwax removal aid.

Urea can also be used as a diuretic. It was first used as a diuretic by a Dr. W. Friedrich in 1892. In a 2010 study of ICU patients in Belgium, urea was used as a diuretic to treat euvolemic hyponatremia and was found a safe, inexpensive, and simple treatment.

Certain types of instant cold packs (or ice packs) contain water and separated urea crystals. Rupturing the internal water bag starts an endothermic reaction that makes the pack cold enough to use to reduce swelling.

Like saline, urea injection is used to perform abortion.

Urea is the main component of an alternative medicinal treatment referred to as urine therapy.

The blood urea nitrogen (BUN) test is a measure of the amount of nitrogen in the blood that comes from urea. It is used as a marker of renal function, though it is inferior to other markers such as creatinine because blood urea levels are influenced by other factors such as diet and dehydration.

Urea labeled with carbon-14 or carbon-13 is used in the urea breath test, which is used to detect the presence of the bacteria *Helicobacter pylori* (*H. pylori*) in the stomach and duodenum of humans, associated with peptic ulcers. The test detects the characteristic enzyme urease, produced by *H. pylori*, by a reaction that produces ammonia from urea. This increases the pH (reduces acidity) of the stomach environment around the bacteria. Similar bacteria species to *H. pylori* can be identified by the same test in animals such as apes, dogs, and cats (including big cats).

Miscellaneous Uses

- A component of animal feed, providing a relatively cheap source of nitrogen to promote growth

- A non-corroding alternative to rock salt for road de-icing, and the hardening of ski-resort terrain park take-offs and landings

- A flavor-enhancing additive for cigarettes

- A main ingredient in hair removers such as Nair and Veet

- A browning agent in factory-produced pretzels

- An ingredient in some skin cream, moisturizers, hair conditioners

- A cloud seeding agent, along with other salts

- A flame-proofing agent, commonly used in dry chemical fire extinguisher charges such as the urea-potassium bicarbonate mixture

- An ingredient in many tooth whitening products

- An ingredient in dish soap

- Along with diammonium phosphate, as a yeast nutrient, for fermentation of sugars into ethanol

- A nutrient used by plankton in ocean nourishment experiments for geoengineering purposes

- As an additive to extend the working temperature and open time of hide glue

- As a solubility-enhancing and moisture-retaining additive to dye baths for textile dyeing or printing

- An ingredient in swimming pool conditioner which prevents UV rays from breaking down the chlorine in the water

Adverse Effects

Urea can be irritating to skin, eyes, and the respiratory tract. Repeated or prolonged contact with urea in fertilizer form on the skin may cause dermatitis.

High concentrations in the blood can be damaging. Ingestion of low concentrations of urea, such as are found in typical human urine, are not dangerous with additional water ingestion within a reasonable time-frame. Many animals (e.g., dogs) have a much more concentrated urine and it contains a higher urea amount than normal human urine; this can prove dangerous as a source of liquids for consumption in a life-threatening situation (such as in a desert).

Urea can cause algal blooms to produce toxins, and its presence in the runoff from fertilized land may play a role in the increase of toxic blooms.

The substance decomposes on heating above melting point, producing toxic gases, and reacts violently with strong oxidants, nitrites, inorganic chlorides, chlorites and perchlorates, causing fire and explosion.

Physiology

Amino acids from ingested food that are not used for the synthesis of proteins and other biological substances — or produced from catabolism of muscle protein — are oxidized by the body as an alternative source of energy, yielding urea and carbon dioxide. The oxidation pathway starts with the removal of the amino group by a transaminase; the amino group is then fed into the urea cycle. The first step in the conversion of amino acids from protein into metabolic waste in the liver is removal of the alpha-amino nitrogen, which results in ammonia. Because ammonia is toxic, it is excreted immediately by fish, converted into uric acid by birds, and converted into urea by mammals.

Ammonia (NH_3) is a common byproduct of the metabolism of nitrogenous compounds. Ammonia is smaller, more volatile and more mobile than urea. If allowed to accumulate, ammonia would raise the pH in cells to toxic levels. Therefore, many organisms convert ammonia to urea, even though this synthesis has a net energy cost. Being practically neutral and highly soluble in water, urea is a safe vehicle for the body to transport and excrete excess nitrogen.

Urea is synthesized in the body of many organisms as part of the urea cycle, either from the oxidation of amino acids or from ammonia. In this cycle, amino groups donated by ammonia and L-aspartate are converted to urea, while L-ornithine, citrulline, L-argininosuccinate, and L-arginine act as intermediates. Urea production occurs in the liver and is regulated by N-acetylglutamate. Urea is then dissolved into the blood (in the reference range of 2.5 to 6.7 mmol/liter) and further transported and excreted by the kidney as a component of urine. In addition, a small amount of urea is excreted (along with sodium chloride and water) in sweat.

In water, the amine groups undergo slow displacement by water molecules, producing ammonia, ammonium ion, and bicarbonate ion. For this reason, old, stale urine has a stronger odor than fresh urine.

Humans

The cycling of and excretion of urea by the kidneys is a vital part of mammalian metabolism. Besides its role as carrier of waste nitrogen, urea also plays a role in the countercurrent exchange system of the nephrons, that allows for re-absorption of water and critical ions from the excreted urine. Urea is reabsorbed in the inner medullary collecting ducts of the nephrons, thus raising the osmolarity in the medullary interstitium surrounding the thin descending limb of the loop of Henle, which makes the water reabsorb.

By action of the urea transporter 2, some of this reabsorbed urea eventually flows back into the thin ascending limb of the tubule, through the collecting ducts, and into the excreted urine. The body uses this mechanism, which is controlled by the antidiuretic hormone, to create hyperosmotic urine—i.e., urine with a higher concentration of dissolved substances than the blood plasma. This mechanism is important to prevent the loss of water, maintain blood pressure, and maintain a suitable concentration of sodium ions in the blood plasma.

The equivalent nitrogen content (in gram) of urea (in mmol) can be estimated by the conversion factor 0.028 g/mmol. Furthermore, 1 gram of nitrogen is roughly equivalent to 6.25 grams of protein, and 1 gram of protein is roughly equivalent to 5 grams of muscle tissue. In situations such as muscle wasting, 1 mmol of excessive urea in the urine (as measured by urine volume in litres multiplied by urea concentration in mmol/l) roughly corresponds to a muscle loss of 0.67 gram.

Other Species

In aquatic organisms the most common form of nitrogen waste is ammonia, whereas land-dwelling organisms convert the toxic ammonia to either urea or uric acid. Urea is found in the urine of mammals and amphibians, as well as some fish. Birds and saurian reptiles have a different form of nitrogen metabolism that requires less water, and leads to nitrogen excretion in the form of uric acid. It is noteworthy that tadpoles excrete ammonia but shift to urea production during metamorphosis. Despite the generalization above, the urea pathway has been documented not only in mammals and amphibians but in many other organisms as well, including birds, invertebrates, insects, plants, yeast, fungi, and even microorganisms.

Analysis

Urea is readily quantified by a number of different methods, such as the diacetyl monoxime colorimetric method, and the Berthelot reaction (after initial conversion of urea to ammonia via urease). These methods are amenable to high throughput instrumentation, such as automated flow injection analyzers and 96-well micro-plate spectrophotometers.

Related Compounds

The term "urea" is also used for a *class* of chemical compounds that share the same functional group, a carbonyl group attached to two organic amine residues: RR'N—CO—NRR'. Examples include carbamide peroxide, allantoin, and hydantoin. Ureas are closely related to biurets and related in structure to amides, carbamates, carbodiimides, and thiocarbamides.

History

Urea was first discovered in urine in 1727 by the Dutch scientist Herman Boerhaave, although this discovery is often attributed to the French chemist Hilaire Rouelle.

Boerhaave used the following steps to isolate urea:

1. Boiled off water, resulting in a substance similar to fresh cream

2. Used filter paper to squeeze out remaining liquid

3. Waited a year for solid to form under an oily liquid

4. Removed the oily liquid

5. Dissolved the solid in water

6. Used recrystallization to tease out the urea

In 1828, the German chemist Friedrich Wöhler obtained urea artificially by treating silver cyanate with ammonium chloride.

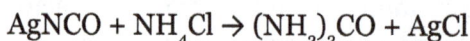

$$AgNCO + NH_4Cl \rightarrow (NH_2)_2CO + AgCl$$

This was the first time an organic compound was artificially synthesized from inorganic starting materials, without the involvement of living organisms. The results of this experiment implicitly discredited vitalism — the theory that the chemicals of living organisms are fundamentally different from those of inanimate matter. This insight was important for the development of organic chemistry. His discovery prompted Wöhler to write triumphantly to Berzelius: "I must tell you that I can make urea without the use of kidneys, either man or dog. Ammonium cyanate is urea." With his discovery, Wöhler secured a place among the pioneers of organic chemistry.

Production

Urea is produced on an industrial scale: In 2012, worldwide production capacity was approximately 184 million tonnes.

Industrial Methods

For use in industry, urea is produced from synthetic ammonia and carbon dioxide. As large quantities of carbon dioxide are produced during the ammonia manufacturing process as a byproduct from hydrocarbons (predominantly natural gas, less often petroleum derivatives), or occasionally from coal, urea production plants are almost always located adjacent to the site where the ammonia is manufactured. Although natural gas is both the most economical and the most widely available ammonia plant feedstock, plants using it do not produce quite as much carbon dioxide from the process as is needed to convert their entire ammonia output into urea. In recent years new technologies such as the KM-CDR process have been developed to recover supplementary carbon dioxide from the combustion exhaust gases produced in the fired reforming furnace of the ammonia synthesis gas plant, allowing operators of stand-alone nitrogen fertilizer complexes to avoid the need to handle and market ammonia as a separate product and also to reduce their greenhouse gas emissions to the atmosphere.

Synthesis

The basic process, developed in 1922, is also called the Bosch–Meiser urea process after its discoverers. Various commercial urea processes are characterized by the conditions under which urea forms and the way that unconverted reactants are further processed. The process consists of two main equilibrium reactions, with incomplete conversion of the reactants. The first is carbamate formation: the fast exothermic reaction of liquid ammonia with gaseous carbon dioxide (CO_2) at high temperature and pressure to form ammonium carbamate ($H_2N\text{-}COONH_4$):

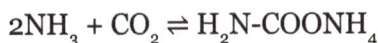

$$2NH_3 + CO_2 \rightleftharpoons H_2N\text{-}COONH_4$$

The second is urea conversion: the slower endothermic decomposition of ammonium carbamate into urea and water:

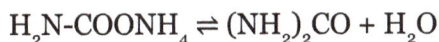

$$H_2N\text{-}COONH_4 \rightleftharpoons (NH_2)_2CO + H_2O$$

Urea plant using ammonium carbamate briquettes, Fixed Nitrogen Research Laboratory, ca. 1930

The overall conversion of NH_3 and CO_2 to urea is exothermic, the reaction heat from the first reaction driving the second. Like all chemical equilibria, these reactions behave according to Le Chatelier's principle, and the conditions that most favour carbamate formation have an unfavourable effect on the urea conversion equilibrium. The process conditions are, therefore, a compromise: the ill-effect on the first reaction of the high temperature (around 190 °C) needed for the second is compensated for by conducting the process under high pressure (140–175 bar), which favours the first reaction. Although it is necessary to compress gaseous carbon dioxide to this pressure, the ammonia is available from the ammonia plant in liquid form, which can be pumped into the system much more economically. To allow the slow urea formation reaction time to reach equilibrium a large reaction space is needed, so the synthesis reactor in a large urea plant tends to be a massive pressure vessel.

Because the urea conversion is incomplete, the product must be separated from unchanged ammonium carbamate. In early "straight-through" urea plants this was done by letting down the system pressure to atmospheric to let the carbamate decompose back to ammonia and carbon dioxide. Originally, because it was not economic to re-compress the ammonia and carbon dioxide for recycle, the ammonia at least would be used for the manufacture of other products, for example ammonium nitrate or sulfate. (The carbon dioxide was usually wasted.) Later process schemes made recycling unused ammonia and carbon dioxide practical. This was accomplished by depressurizing the reaction solution in stages (first to 18–25 bar and then to 2–5 bar) and passing it at each stage through a steam-heated *carbamate decomposer*, then recombining the resultant carbon dioxide and ammonia in a falling-film *carbamate condenser* and pumping the carbamate solution into the previous stage.

The Stripping Concept

The "total recycle" concept has two main disadvantages. The first is the complexity of the flow scheme and, consequently, the amount of process equipment needed. The second is the amount of water recycled in the carbamate solution, which has an adverse effect on the equilibrium in the urea conversion reaction and thus on overall plant efficiency. The stripping concept, developed in the early 1960s by Stamicarbon in The Netherlands, addressed both problems. It also improved heat recovery and reuse in the process.

The position of the equilibrium in the carbamate formation/decomposition depends on the product of the partial pressures of the reactants. In the total recycle processes, carbamate decomposition is promoted by reducing the overall pressure, which reduces the partial pressure of both ammonia and carbon dioxide. It is possible, however, to achieve a similar effect without lowering the overall pressure—by suppressing the partial pressure of just one of the reactants. Instead of feeding carbon dioxide gas directly to the reactor with the ammonia, as in the total recycle process, the stripping process first routes the carbon dioxide through a stripper (a carbamate decomposer that operates under full system pressure and configured to provide maximum gas-liquid contact). This flushes out free ammonia, reducing its partial pressure over the liquid surface and carrying it directly to a carbamate condenser (also under full system pressure). From there, reconstituted ammonium carbamate liquor passes directly to the reactor. That eliminates the medium-pressure stage of the total recycle process altogether.

The stripping concept was such a major advance that competitors such as Snamprogetti—now Saipem—(Italy), the former Montedison (Italy), Toyo Engineering Corporation (Japan), and Urea Casale (Switzerland) all developed versions of it. Today, effectively all new urea plants use the principle, and many total recycle urea plants have converted to a stripping process. No one has proposed a radical alternative to the approach. The main thrust of technological development today, in response to industry demands for ever larger individual plants, is directed at re-configuring and re-orientating major items in the plant to reduce size and overall height of the plant, and at meeting challenging environmental performance targets.

Side Reactions

It is fortunate that the urea conversion reaction is slow—because if it were not it would go into reverse in the stripper. As it is, succeeding stages of the process must be designed to minimize residence times, at least until the temperature reduces to the point where the reversion reaction is very slow.

Two reactions produce impurities. Biuret is formed when two molecules of urea combine with the loss of a molecule of ammonia.

$$2NH_2CONH_2 \rightarrow H_2NCONHCONH_2 + NH_3$$

Normally this reaction is suppressed in the synthesis reactor by maintaining an excess of ammonia, but after the stripper, it occurs until the temperature is reduced. Biuret is undesirable in fertilizer urea because it is toxic to crop plants, although to what extent depends on the nature of the crop and the method of application of the urea. (Biuret is actually welcome in urea when is used as a cattle feed supplement.)

Isocyanic acid results from the thermal decomposition of ammonium cyanate, which is in chemical equilibrium with urea:

$$NH_2CONH_2 \rightarrow NH_4NCO \rightarrow HNCO + NH_3$$

This reaction is at its worst when the urea solution is heated at low pressure, which happens when the solution is concentrated for prilling or granulation. The reaction products mostly volatilize into the overhead vapours, and recombine when these condense to form urea again, which contaminates the process condensate.

Corrosion

Ammonium carbamate solutions are notoriously corrosive to metallic construction materials, even more resistant forms of stainless steel—especially in the hottest parts of the plant such as the stripper. Traditionally corrosion has been minimized (although not eliminated) by continuously injecting a small amount of oxygen (as air) into the plant to establish and maintain a passive oxide layer on exposed stainless steel surfaces. Because the carbon dioxide feed is recovered from ammonia synthesis gas, it contains traces of hydrogen that can mingle with passivation air to form an explosive mixture if allowed to accumulate.

In the mid 1990s two duplex (ferritic-austenitic) stainless steels were introduced (DP28W, jointly developed by Toyo Engineering and Sumitomo Metals Industries and Safurex, jointly developed by Stamicarbon and Sandvik Materials Technology (Sweden).) These let manufactures drastically reduce the amount of passivation oxygen. In theory, they could operate with no oxygen.

Saipem now uses either zirconium stripper tubes, or bimetallic tubes with a titanium body (cheaper but less erosion-resistant) and a metallurgically bonded internal zirconium lining. These tubes are fabricated by ATI Wah Chang (USA) using its Omegabond technique.

Finishing

Urea can be produced as prills, granules, pellets, crystals, and solutions.

Solid Forms

For its main use as a fertilizer urea is mostly marketed in solid form, either as prills or granules. The advantage of prills is that, in general, they can be produced more cheaply than granules and that the technique was firmly established in industrial practice long before a satisfactory urea granulation process was commercialized. However, on account of the limited size of particles that can be produced with the desired degree of sphericity and their low crushing and impact strength, the performance of prills during bulk storage, handling and use is generally (with some exceptions) considered inferior to that of granules.

High-quality compound fertilizers containing nitrogen co-granulated with other components such as phosphates have been produced routinely since the beginnings of the modern fertilizer industry, but on account of the low melting point and hygroscopic nature of urea it took courage to apply the same kind of technology to granulate urea on its own. But at the end of the 1970s three companies began to develop fluidized-bed granulation. The first in the field was Nederlandse Stikstof Maatschappij, which later became part of Hydro Agri (now Yara International). Yara eventually sold this technology to Uhde GmbH, whose Uhde Fertilizer Technology (UFT) subsidiary now markets it. Around the same time Toyo Engineering Corporation developed its spouted-bed process, comprising a fluidized bed deliberately agitated to produce turbulent ebullation. Stamicarbon also undertook development work on its own fluidized-bed granulation system, using film sprays rather than atomizing sprays to introduce the urea melt, but shelved it until the 1990s, when there was for a time considerable doubt about the commercial future of the Hydro (UFT) process. As a result, the Stamicarbon technology is now commercialized and highly successful. More recently, Urea Casale introduced a different fluidized-bed granulation system: the urea is sprayed in laterally from the side walls of the granulator instead of from the bottom. This organizes the bed into two cylindrical masses contrarotating on parallel longitudinal axes. The raw product is uniform enough to not require screens.

Surprisingly, perhaps, considering the product particles not spherical, pastillation using a Rotoform steel-belt pastillator is gaining ground as a urea particle-forming process as a result of development work by Stamicarbon in collaboration with Sandvik Process Systems (Germany). Single-machine capacity is limited to 175 t/d, but the machines are simple and need little maintenance, specific power consumption is much lower than for granulation, and the product is very uniform. The robustness of the product appears to make up for its non-spherical shape.

UAN Solutions

In admixture, the combined solubility of ammonium nitrate and urea is so much higher than that of either component alone that it is possible to obtain a stable solution (known as UAN) with a total nitrogen content (32%) approaching that of solid ammonium nitrate (33.5%), though not, of course, that of urea itself (46%). Given the ongoing safety and security concerns surrounding fertilizer-grade solid ammonium nitrate, UAN provides a considerably safer alternative without entirely sacrificing the agronomic properties that make ammonium nitrate more attractive than urea as a fertilizer in areas with short growing seasons. It is also more convenient to store and handle than a solid product and easier to apply accurately to the land by mechanical means.

Laboratory Preparation

Ureas in the more general sense can be accessed in the laboratory by reaction of phosgene with primary or secondary amines, proceeding through an isocyanate intermediate. Non-symmetric ureas can be accessed by reaction of primary or secondary amines with an isocyanate.

Also, urea is produced when phosgene reacts with ammonia:

$$COCl_2 + 4\,NH_3 \rightarrow (NH_2)_2CO + 2\,NH_4Cl$$

Urea is byproduct of converting alkyl halides to thiols via a S-alkylation of thiourea. Such reactions proceed via the intermediacy of isothiouronium salts:

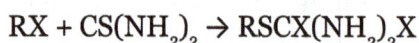

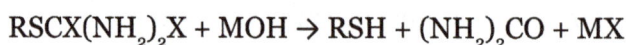

$$RX + CS(NH_2)_2 \rightarrow RSCX(NH_2)_2X$$

$$RSCX(NH_2)_2X + MOH \rightarrow RSH + (NH_2)_2CO + MX$$

In this reaction R is alkyl group, X is halogen and M is alkali metal.

Historical Process

Urea was first noticed by Hermann Boerhaave in the early 18th century from evaporates of urine. In 1773, Hilaire Rouelle obtained crystals containing urea from human urine by evaporating it and treating it with alcohol in successive filtrations. This method was aided by Carl Wilhelm Scheele's discovery that urine treated by concentrated nitric acid precipitated crystals. Antoine François, comte de Fourcroy and Louis Nicolas Vauquelin discovered in 1799 that the nitrated crystals were identical to Rouelle's substance and invented the term "urea." Berzelius made further improvements to its purification and finally William Prout, in 1817, succeeded in obtaining and determining the chemical composition of the pure substance. In the evolved procedure, urea was precipitated as urea nitrate by adding strong nitric acid to urine. To purify the resulting crystals, they were dissolved in boiling water with charcoal and filtered. After cooling, pure crystals of urea nitrate form. To reconstitute the urea from the nitrate, the crystals are dissolved in warm water, and barium carbonate added. The water is then evaporated and anhydrous alcohol added to extract the urea. This solution is drained off and evaporated, leaving pure urea.

Chemical Properties

Molecular and Crystal Structure

The urea molecule is planar in the crystal structure, but the geometry around the nitrogens is pyramidal in the gas-phase minimum-energy structure. In solid urea, the oxygen center is engaged in two N-H-O hydrogen bonds. The resulting dense and energetically favourable hydrogen-bond network is probably established at the cost of efficient molecular packing: The structure is quite open, the ribbons forming tunnels with square cross-section. The carbon in urea is described as sp² hybridized, the C-N bonds have significant double bond character, and the carbonyl oxygen is basic compared to, say, formaldehyde. Urea's high aqueous solubility reflects its ability to engage in extensive hydrogen bonding with water.

By virtue of its tendency to form porous frameworks, urea has the ability to trap many organic compounds. In these so-called clathrates, the organic "guest" molecules are held in channels formed by interpenetrating helices composed of hydrogen-bonded urea molecules. This behaviour can be used to separate mixtures, e.g., in the production of aviation fuel and lubricating oils, and in the separation of hydrocarbons.

As the helices are interconnected, all helices in a crystal must have the same molecular handedness. This is determined when the crystal is nucleated and can thus be forced by seeding. The resulting crystals have been used to separate racemic mixtures.

Reactions

Urea reacts with alcohols to form urethanes. Urea reacts with malonic esters to make barbituric acids.

Coated Urea

Coated urea fertilizers are a group of controlled release fertilizers consisting of prills of urea coated in less-soluble chemicals such as sulfur, polymers, other products or a combination. These fertilizers mitigate some of the negative aspects of urea fertilization, such as fertilizer burn. The coatings release the urea either when penetrated by water, as with sulfur, or when broken down, as with polymers.

Overview

Urea is widely used as a nitrogen fertilizer. Its high solubility in water makes it useful for liquid application, and it has a much lower risk of causing fertilizer burn than other chemicals such as calcium cyanide or ammonium nitrate. However, the risk of fertilizer burn with urea can be unacceptably high in some situations, such as higher temperatures. The high water-solubility of urea can be disadvantageous in some cases as well.

One particular technique to mitigate these disadvantages has been to encapsulate prills of urea with less-soluble chemicals. These coatings permit the gradual release of urea in a controlled fashion, allowing for less-frequent applications.

Sulfur-coated Urea

Sulfur-coated urea, or SCU, fertilizers release nitrogen via water penetration through cracks and micropores in the coating. Once water penetrates through the coating, nitrogen release is rapid. The particles of fertilizer may in turn be sealed with wax to slow release further still, making microbial degradation necessary to permit water penetration. The size of fertilizer particles may also be varied in order to vary the time at which nitrogen release occurs. Sulfur-coated products typically range from 32% to 41% elemental nitrogen by weight. The sulfur coating process was originally developed by the Tennessee Valley Authority.

Sulfur-coated urea products can only be applied in granular form, and thus cannot be applied via liquid fertilization methods. It is not uncommon to find empty sulfur husks in turf once the nitrogen is released. Another disadvantage has to do with the relatively large particle size of sulfur-coated urea fertilizers, which makes their use on closely mown surfaces like putting greens impractical. However, more recently, materials with smaller particle sizes have been developed, permitting the use of sulfur-coated ureas on putting greens.

Polymer-Coated Urea

Polymer-coated urea, also called plastic-coated urea, or PCU, fertilizers can permit a more precise rate of nitrogen release than sulfur-coated urea products. It's possible to produce polymer-coated products where the nitrogen release can be delayed for 10 months after application. The primary disadvantage of polymer-coated urea products is their relatively high cost compared to sulfur-coated urea.

Combination Products

Products that use a combination of sulfur-coating and polymer-coating also exist. Typically, these products consist of urea, coated with a layer of sulfur, which is in turn coated with a layer of polymer. Each coating layer is generally less than the normal thickness for the individual processes. These products are generally used as less-expensive alternatives to purely plastic-coated products, while still providing precise nitrogen release characteristics.

Urea-formaldehyde

Urea-formaldehyde, also known as urea-methanal, so named for its common synthesis pathway and overall structure, is a non-transparent thermosetting resin or plastic, made from urea and formaldehyde heated in the presence of a base. These resins are used in adhesives, finishes, particle board, MDF, and molded objects. UF and related amino resins are considered a class of thermosetting resins of which urea-formaldehyde resins make up 80% produced globally. Examples of amino resins use include in automobile tires to improve the bonding of rubber to tire cord, in paper for improving tear strength, in molding electrical devices, jar caps, etc.

Properties

Urea-formaldehyde resin's attributes include high tensile strength, flexural modulus, and a high heat distortion temperature, low water absorption, mould shrinkage, high surface hardness, elongation at break, and volume resistance.

Index of Refraction = 1.55

Chemical Structure

The chemical structure of UF resins can be described as that of an aldehyde condensation polymer. This description leaves the details of the structure undetermined, which can vary linearly and branched. These are grouped by their average molar mass and the content of different functional groups. Changing synthesis conditions of the resins give good designing possibilities for the structure and resin properties. Generally the polymer is prepared by reacting the urea and formaldehyde in a 1:1 ratio. The result is a polymer with NCH_2N repeat units, not NCH_2OCH_2N repeat units found in melanine-formaldehyde.

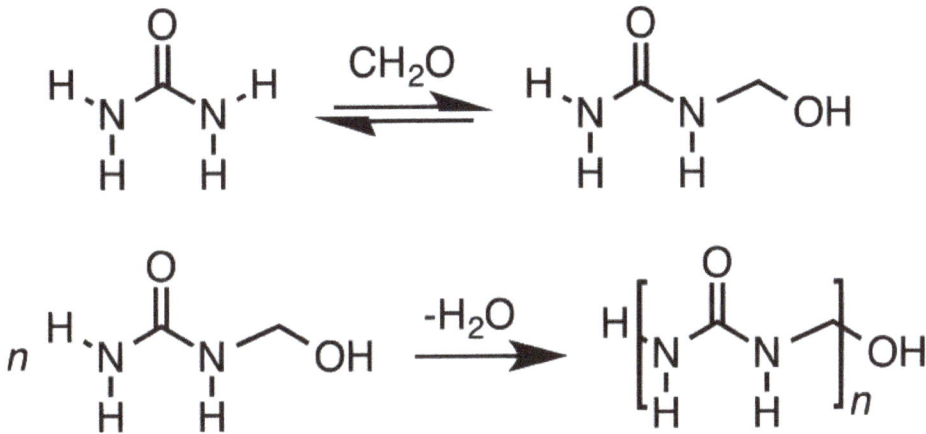

Two steps in formation of urea-formaldehyde resin

Production

Approximately 1 million metric tons of urea-formaldehyde are produced every year. Over 70% of this production is then put into use by the forest products industry for bonding particleboard (61%), medium density fiberboard (27%), hardwood plywood (5%), and laminating adhesive (7%).

General uses

Urea-formaldehyde is everywhere and used in many manufacturing processes due to its useful properties. Examples include decorative laminates, textiles, paper, foundry sand molds, wrinkle resistant fabrics, cotton blends, rayon, corduroy, etc. It is also used to glue wood together. Urea formaldehyde was commonly used when producing electrical appliances casing (e.g. desk lamps).

A range of objects made from urea formaldehyde

Agricultural use

Urea formaldehyde is also used in agriculture as a controlled release source of nitrogen fertilizer. Urea formaldehyde's rate of decomposition into CO_2 and NH_3 is determined by the action of microbes found naturally in most soils. The activity of these microbes, and, therefore, the rate of nitrogen release, is temperature dependent. The optimum temperature for microbe activity is approximately 70-90 °F (approx 20-30 °C).

Urea-formaldehyde Foam Insulation

Urea-formaldehyde insulation

Urea-formaldehyde foam insulation (UFFI) dates to the 1930s and made a synthetic insulation with R-values near 5.0 per inch. It is basically a foam, like shaving cream, that is easily injected into walls with a hose. It is made by using a pump set and hose with a mixing gun to mix the foaming agent, resin and compressed air. The fully expanded foam is pumped into areas in need of insulation. It becomes firm within minutes but cures within a week. UFFI is generally found in homes built before the 1970s, often in basements, crawl spaces, attics, and unfinished attics. Visually it looks like oozing liquid that has been hardened. Over time, it tends to vary in shades of butterscotch but new UFFI is a light yellow color. Early forms of UFFI tended to shrink significantly. Modern UF insulation with updated catalysts and foaming technology have reduced shrinkage to minimal levels (between 2-4%). The foam dries with a dull matte color with no shine. When cured, it often has a dry and crumbly texture.

Safety Concerns

Urea-formaldehyde foam insulation (UFFI) was used extensively in the 1970s. Home-owners used UFFI as a wall cavity filler at the time in order to conserve energy. In the 1980s, concerns began to develop about formaldehyde vapor emitted in the curing process, as well as from the breakdown of old foam. Emission rates exceeding 3.0 - 5.0 parts per million (ppm) cause a variety of adverse health effects impacting the eyes, nose, and respiratory system. Consequently, its use was discontinued. The urea-form-aldehyde emissions decline over time and significant levels should no longer be present in the homes today. Modern replacement options for UFFI include melamine form-aldehyde resin, low-emission UF insulation materials, and polyurethane. The curing temperature is unknown.

UFFI was usually mixed at the location of use while constructing the home's walls. It was then injected inside the walls, the curing process occurs, and the final product acts as an insulating agent. Because less information was known about the toxic health effects of formaldehyde in the 1970s, extra formaldehyde was often added to the mix-ture to ensure that the curing process would occur completely. Since the UFFI was not a well-sealed product [open-celled foam], any excess formaldehyde in the insulation would off-gas into the home's living space. The early UFFI materials were also affected by moisture and heat which compounded the offgassing concerns. When temperature rises, residuals of formaldehyde contained in the insulation are released and migrate into indoor air. Remedial actions to take when formaldehyde levels exceed recommend-ed levels include sealing off the any outlets for the vapors; sealing any cracks or open-ings in interior walls; removing any sources of water or moisture that come in contact with the insulation; applying one or more layers of vapor-barrier paint; increasing the air exchange rate with outside air in buildings that are tightly sealed; or covering walls with Mylar or vinyl paper. Aluminum foil is a useful alternative for barricading vapors. Generally there is not an off-gassing concern with older UFFI insulation, since those materials have already cured. Removal is a costly and tedious option for UFFI, and it requires the installation of replacement insulation.

Health Concerns

Health effects occur when urea-formaldehyde based materials and products release formaldehyde into the air. Generally there are no observable health effects from form-aldehyde when air concentrations are below 1.0 ppm. The onset of respiratory irritation and other health effects, and even increased cancer risk begins when air concentrations exceed 3.0-5.0 ppm. This triggers watery eyes, nose irritations, wheezing and coughing, fatigue, skin rash, severe allergic reactions, burning sensations in the eyes and throat, nausea, and difficulty in breathing in some humans (usually > 1.0 ppm). Occupants of

UFFI insulated homes with elevated formaldehyde levels experienced systemic symptoms such as headache, malaise, insomnia, anorexia, and loss of libido. Irritation of the mucous membranes (specifically the eyes, nose, and throat) was a common upper respiratory tract symptom related to formaldehyde exposure. However, when compared to control groups, the frequency of symptoms did not exceed the controls except when it came to wheezing, difficult breathing, and a burning skin sensation. Controlled studies have suggested that tolerance to formaldehyde's odor and irritating effects can occur over a prolonged exposure.

History

It was first synthesized in 1884 by Hölzer, who was working with Bernhard Tollens. In 1919, Hanns John (1891–1942) of Prague, Czechoslovakia obtained the first patent for urea-formaldehyde resin.

Excellerator (Brand)

Excellerator	
Product type	Specialty micronutrient fertilizer
Owner	Harsco Minerals
Country	United States
Markets	United States
Website	http://www.harscominerals.com/products/product-line.aspx?id=1103

Excellerator is a specialty micronutrient fertilizer produced by the U.S.-based company Harsco Minerals. It is a granular pelletized product used on golf courses, athletic fields and in the lawn and garden market. Excellerator aides in the correction of plant and soil nutrient imbalances and metal toxicities. It provides high concentrations of plant-available silicon which has been shown in university and field trials to enhance plant resistance to biological and environmental stresses and improve plant nutrient uptake.

How It's Made

Excellerator is a co-product of downstream stainless steel furnace slag that undergoes proprietary aging processes, fine grind pulverizing, and extensive metal separation to remove impurities. Over 99.95% of all metals are removed. Particular attention is paid to the removal of "free lime" (CaO and MgO), which disrupts product stability and consistency of performance. The final product contains both calcium and magnesium silicates in addition to a micronutrient package.

Micronutrient Analysis

(Derived from calcium and magnesium silicates, boric acid, zinc sulfate, and copper sulfate)

Calcium (Ca)	24.00%
Magnesium (Mg)	6.00%
Boron (B)	0.02%
Copper (Cu)	0.05%
Iron (Fe)	1.80%
Manganese (Mn)	0.50%
Molybdenum (Mo)	0.002%
Zinc (Zn)	0.05%

Silicon as a Beneficial Nutrient

The amount of total silicon in Excellerator is 39%. Silicon has been shown to improve plant cell wall strength and structural integrity in many turfgrass species. Other benefits of silicon to plants include increased drought and frost resistance, decreased lodging and improved plant response to pests and disease. Silicon has also been shown to improve plant vigor and physiology, resulting in increases to plant root and above ground biomass. For turfgrass, these benefits may result in better turf quality, color, density, and wear tolerance. Silicon has also been shown effective in enhancing the suppression of diseases, such as grey leaf spot, in a number of warm and cool season turfgrass species.

Silicon is a naturally occurring mineral and the second most abundant element in the earth's crust. In the soil, silicon attaches to soil colloids, helping to reduce compaction and making tied-up nutrients more available. Silicon also allows for a faster, more efficient movement of calcium and magnesium through the soil and readily ties up toxic elements, like aluminium, reducing metal toxicity. In the plant, silicon stengthens cell walls; improving plant strength, health, and productivity. Although not considered an essential element for plant growth and development, silicon is considered a beneficial element in many countries throughout the world due to its many benefits to numerous plant species when under abiotic or biotic stresses. Silicon is currently under consideration by the Association of American Plant Food Control Officials (AAPFCO) for elevation to the status of a "plant beneficial substance."

Labeling of Fertilizer

The labeling of fertilizers varies by country in terms of analysis methodology, nutrient labeling, and minimum nutrient requirements. The most common labeling convention shows the amounts of nitrogen, phosphorus, and potassium in the fertilizer.

Labeling of Macronutrient Fertilizers

Macronutrient fertilizers are generally labeled with an *NPK* analysis, based on the relative content of the chemical elements nitrogen (N), phosphorus (P), and potassium (K) that are commonly used in fertilizers. However, numbers used in this labeling scheme do not directly represent the source composition or absolute nutrient content of the fertilizer. The N value is the percentage of elemental nitrogen by weight in the fertilizer. The value for P is the fraction by weight of P_2O_5 in a fertilizer with the same amount of phosphorus that gets all of its phosphorus from P_2O_5. The value for K is analogous, based on a fertilizer with K_2O.

For example, the fertilizer *sylvite* is a naturally occurring mineral consisting mostly of potassium chloride (*KCl*). As such, it contains one potassium atom for every chlorine atom, and is 52% potassium and 48% chlorine *by weight* (because of the difference in atomic weights of the elements). K_2O is similarly 83% potassium. To have 52% potassium, therefore, a fertilizer that gets all its potassium from K_2O would have to be 63% K_2O (.52/.83 is .63). Pure KCl fertilizer would thus be labeled 0-0-63; because sylvite is less than pure (it contains other compounds that contain no potassium), it is labeled 0-0-60.

Converting Nutrient Analysis to Composition

The factors for converting from P_2O_5 and K_2O values on a fertilizer label to the concentrations (by weight) of P and K elements are as follows:

- P_2O_5 consists of 56.4% oxygen and 43.6% elemental phosphorus. The percentage (mass fraction) of elemental phosphorus is 43.6% so elemental P = 0.436 x P_2O_5

- K_2O consists of 17% oxygen and 83% elemental potassium. The percentage (mass fraction) of elemental potassium is 83% so elemental K = 0.83 x K_2O

- Nitrogen values represent actual nitrogen content so these numbers do not need to be converted.

Using these conversion factors, an 18–51–20 fertilizer contains by weight:

- 18% elemental (N)

- 22% elemental (P), and

- 17% elemental (K)

Other Labeling Conventions

In the U.K., fertilizer labeling regulations allow for reporting the elemental mass fractions of phosphorus and potassium. The regulations stipulate that this should be done in parentheses after the standard N-P-K values. In Australia, macronutrient fertilizers are labeled with an "N-P-K-S" system, which uses elemental mass fractions rather than standard N-P-K values and includes the amount of sulfur (S) contained in the fertilizer.

NPK Values for Commercial Fertilizers

NPK Values for Various Synthetic Fertilizers

- 15-00-00 Calcium nitrate

- 21-00-00 Ammonium sulfate

- 30-00-00 to 40-00-00 Sulfur-coated urea (slow release)

- 31-00-00 Isobutylidene diurea (~90% slow release)

- 33-00-00 to 34-00-00 Ammonium nitrate

- 35-00-00 Ureaform (~85% slow release, sparingly soluble ureaformaldehyde)

- 40-00-00 Methylene ureas (~70% slow release)

- 46-00-00 Urea (U-46)

- 82-00-00 Anhydrous ammonia

- 10-34-00 to 11-37-00 Ammonium polyphosphate

- 11-48-00 to 11-55-00 Monoammonium phosphate

- 18-46-00 to 21-54-00 Diammonium phosphate

- 13-00-44 Potassium nitrate

- 00-17-00 to 00-22-00 Superphosphate (Monocalcium phosphate monohydrate with gypsum)

- 00-44-00 to 00-52-00 Triple superphosphate (Monocalcium phosphate monohydrate)

NPK Values for Mined Fertilizer Minerals

- 11-08-02 to 16-12-03 bird guano

- 00-3-00 to 00-8-00 Raw Phosphate Rock (would be 00-34-00 if it were soluble)

- 00-00-22 Potassium magnesium sulfate (K-mag)

- 00-00-60 Potassium chloride

NPK Values for Biosolids Fertilizers and Others

- 09-00-00 dairy manure

- 01-00-01 horse manure

- 03-02-02 poultry manure

- 04-12-00 Bone meal

- 05-05-06 Fish blood and bone

- 06-02-00 Milorganite

History of Fertilizer

The history of fertilizer has largely shaped political, economic, and social circumstances in their traditional uses. Subsequently, there has been a radical reshaping of environmental conditions following the development of chemically synthesized fertilizers.

Ancient History

Egyptians, Romans, Babylonians, and early Germans all are recorded as using minerals and or manure to enhance the productivity of their farms. The use of wood ash as a field treatment became widespread. In the 1800s Humboldt recommended the use of guano.

Key Figures in Europe

In the 1730s, Viscount Charles Townshend (1674-1738) first studied the improving effects of the four crop rotation system that he had observed in use in Flanders. For this he gained the nickname of Turnip Townshend.

Johann Friedrich Mayer

Johann Friedrich Mayer (1719-1798) was the first to present to the world a series of experiments upon it the relation of gypsum to agriculture, and many chemists have followed him in the 19th century. Early 19th century however a great variety of opinion remained with regard to its mode of operation, for example:

- The French agronomist Victor Yvart (1763-1831) believed that the action of gypsum is exclusively the effect of the sulphuric acid, which enters into its composition; and founds this opinion upon the fact that the ashes of turf, which contain sulphate of iron and sulphate of alumina, have the same action upon vegetation as gypsum.

- The French agronomist Charles Philibert de Lasteyrie (1759-1849), observing that plants whose roots were nearest the surface of the soil were most acted upon by plaster, concludes that gypsum takes from the atmosphere the elements of vegetable life, and transmits them directly to plants.

- Louis Augustin Guillaume Bosc intimates that the septic quality of gypsum (which he takes for granted) best explains its action on vegetation; but this opinion is subverted by the experiments of Davy.

- Humphry Davy found that, of two parcels of minced veal, the one mixed with gypsum, the other left by itself, and both exposed to the action of the sun, the latter was the first to exhibit symptoms of putrefaction. Davy's own belief on this subject is, that it makes part of the food of vegetables, is received into the plant, and combined with it.

Mayer also promote new regimes of crop rotation.

Justus Von Liebig

Chemist Justus von Liebig (1803-1873) contributed greatly to the advancement in the understanding of plant nutrition. His influential works first denounced the vitalist theory of humus, arguing first the importance of ammonia, and later promoting the importance of inorganic minerals to plant nutrition. Primarily Liebig's work succeeded in exposition of questions for agricultural science to address over the next 50 years.

In England, he attempted to implement his theories commercially through a fertilizer created by treating phosphate of lime in bone meal with sulfuric acid. Although it was much less expensive than the guano that was used at the time, it failed because it was not able to be properly absorbed by crops.

Sir John Bennet Lawes

John Bennet Lawes, an English entrepreneur, (view timeline of his life and work) began to experiment on the effects of various manures on plants growing in pots in 1837, and a year or two later the experiments were extended to crops in the field. One immediate consequence was that in 1842 he patented a manure formed by treating phosphates with sulphuric acid, and thus was the first to create the artificial manure industry. In the succeeding year he enlisted the services of Joseph Henry Gilbert, who had studied under Liebig at the University of Giessen, as director of research at the Rothamsted Experimental Station which he founded on his estate. To this day, the Rothamsted research station the pair founded still investigates the impact of inorganic and organic fertilizers on crop yields.

Jean Baptiste Boussingault

In France, Jean Baptiste Boussingault (1802-1887) pointed out that the amount of nitrogen in various kinds of fertilizers is important.

Metallurgists Percy Gilchrist (1851-1935) and Sidney Gilchrist Thomas (1850-1885) invented the Gilchrist-Thomas process, which enabled the use of high phosphorus acidic Continental ores for steelmaking. The dolomite lime lining of the converter turned in time into calcium phosphate, which could be used as fertilizer, known as Thomas-phosphate.

The Birkeland-Eyde Process

The Birkeland–Eyde process was developed by Norwegian industrialist and scientist Kristian Birkeland along with his business partner Sam Eyde in 1903, based on a method used by Henry Cavendish in 1784. This process was used to fix atmospheric nitrogen (N_2) into nitric acid (HNO_3), one of several chemical processes generally referred to as nitrogen fixation. The resultant nitric acid was then used for the production of synthetic fertilizer. A factory based on the process was built in Rjukan and Notodden in Norway, combined with the building of large hydroelectric power facilities. The process is inefficient in terms of energy usage, and is today replaced by the Haber process.

The Haber Process

In the early decades of the 20th century, the Nobel prize-winning chemists Carl Bosch of IG Farben and Fritz Haber developed the Haber process which utilized molecular nitrogen (N_2) and methane (CH_4) gas in an economically sustainable synthesis of ammonia (NH_3). The ammonia produced in the Haber process is the main raw material of the Ostwald process.

The Ostwald process

The Ostwald process is a chemical process for production of nitric acid (HNO_3), which was developed by Wilhelm Ostwald (patented 1902). It is a mainstay of the modern chemical industry and provides the raw material for the most common type of fertilizer production, globally. Historically and practically it is closely associated with the Haber process, which provides the requisite raw material, ammonia (NH_3).

Erling Johnson

In 1927 Erling Johnson developed an industrial method for producing nitrophosphate, also known as the Odda process after his Odda Smelteverk of Norway. The process involved acidifying phosphate rock (from Nauru and Banaba Islands in the southern

Pacific Ocean) with nitric acid to produce phosphoric acid and calcium nitrate which, once neutralized, could be used as a nitrogen fertilizer.

Industry

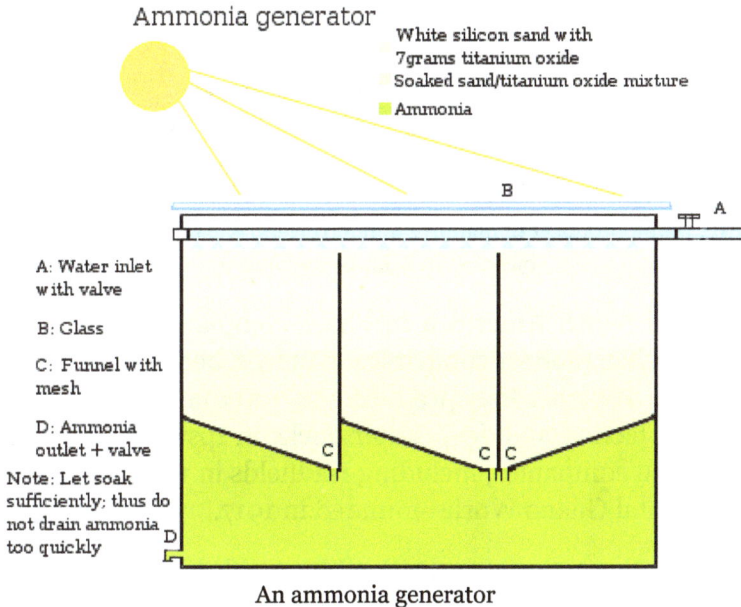

An ammonia generator

British

The Englishmen James Fison, Edward Packard, Thomas Hadfield and the Prentice brothers each founded companies in the early 19th century to create fertilizers from bone meal.

The developing sciences of chemistry and Paleontology, combined with the discovery of coprolites in commercial quantities in East Anglia, led Fisons and Packard to develop sulfuric acid and fertilizer plants at Bramford, and Snape, Suffolk in the 1850s to create superphosphates, which were shipped around the world from the port at Ipswich. By 1871 there were about 80 factories making superphosphate.

After World War I these businesses came under competitive pressure from naturally produced guano, primarily found on the Pacific islands, as their extraction and distribution had become economically attractive.

The interwar period saw innovative competition from Imperial Chemical Industries who developed synthetic ammonium sulfate in 1923, Nitro-chalk in 1927, and a more concentrated and economical fertilizer called CCF (Concentrated Complete Fertiliser) based on ammonium phosphate in 1931. Competition was limited as ICI ensured it controlled most of the world's ammonium sulfate supplies.

North America and Other European Countries

Founded in 1812, Mirat, producer of manures and fertilizers, is claimed to be the oldest industrial business in Salamanca (Spain).

Other European and North American fertilizer companies developed their market share, forcing the English pioneer companies to merge, becoming Fisons, Packard, and Prentice Ltd. in 1929. Together they produced 85,000 tons of superphosphate/year in 1934 from their new factory and deep-water docks in Ipswich. By World War II they had acquired about 40 companies, including Hadfields in 1935, and two years later the large Anglo-Continental Guano Works, founded in 1917.

The post-war environment was characterized by much higher production levels as a result of the "Green Revolution" and new types of seed with increased nitrogen-absorbing potential, notably the high-response varieties of maize, wheat, and rice. This has accompanied the development of strong national competition, accusations of cartels and supply monopolies, and ultimately another wave of mergers and acquisitions. The original names no longer exist other than as holding companies or brand names: Fisons and ICI agrochemicals are part of today's Yara International and AstraZeneca companies.

Major players in this market now include the Russian fertilizer company Uralkali (listed on the London Stock Exchange), whose former majority owner is Dmitry Rybolovlev, ranked by Forbes as 60th in the list of wealthiest people in 2008.

References

- Francis Borgio J, Sahayaraj K and Alper Susurluk I (eds) . Microbial Insecticides: Principles and Applications, Nova Publishers, USA. 492pp. ISBN 978-1-61209-223-2

- Gericke, William F. (1940). The Complete Guide to Soilless Gardening (1st ed.). London: Putnam. pp. 9–10, 38 & 84. ISBN 9781163140499.

- Douglas, James Sholto (1975). Hydroponics: The Bengal System (5th ed.). New Dehli: Oxford University Press. p. 10. ISBN 9780195605662.

- Sholto Douglas, James (1985). Advanced guide to hydroponics: (soiless cultivation). London: Pelham Books. pp. 169–187, 289–320, & 345–351. ISBN 9780720715712.

- J. Benton, Jones (2004). Hydroponics: A Practical Guide for the Soilless Grower (2nd ed.).

Newyork: Taylor & Francis. pp. 29–70 & 225–229. ISBN 9780849331671.

- Quastel, J. H. (1950). "2,4-Dichlorophenoxyacetic Acid (2,4-D) as a Selective Herbicide". Agricultural Control Chemicals. Advances in Chemistry. 1. p. 244. doi:10.1021/ba-1950-0001. ch045. ISBN 0-8412-2442-0.

- Smith (18 July 1995). "8: Fate of herbicides in the environment". Handbook of Weed Management Systems. CRC Press. pp. 245–278. ISBN 978-0-8247-9547-4.

- Powles, S. B.; Shaner, D. L., eds. (2001). Herbicide Resistance and World Grains. CRC Press, Boca Raton, FL. p. 328. ISBN 9781420039085.

- Moss, S. R. (2002). "Herbicide-Resistant Weeds". In Naylor,, R. E. L. Weed management handbook (9th ed.). Blackwell Science Ltd. pp. 225–252. ISBN 0-632-05732-7.

- Oates, J. A. H. (11 July 2008). Lime and Limestone: Chemistry and Technology, Production and Uses. John Wiley & Sons. pp. 111–3. ISBN 978-3-527-61201-7.

- H.A. Mills; J.B. Jones Jr. (1996). Plant Analysis Handbook II: A practical Sampling, Preparation, Analysis, and Interpretation Guide. ISBN 1-878148-05-2.

- J. Benton Jones, Jr. "Inorganic Chemical Fertilizers and Their Properties" in Plant Nutrition and Soil Fertility Manual, Second Edition. CRC Press, 2012. ISBN 978-1-4398-1609-7. eBook ISBN 978-1-4398-1610-3.

- Smil, Vaclav (2004). Enriching the Earth. Massachusetts Institute of Technology. p. 135. ISBN 9780262693134.

- Carroll and Salt, Steven B. and Steven D. (2004). Ecology for Gardeners. Cambridge: Timber Press. ISBN 9780881926118.

- Appl, Max (2000). Ullmann's Encyclopedia of Industrial Chemistry, Volume 3. Weinheim, Germany: Wiley-VCH Verlag GmbH & Co. KGaA. pp. 139–225. doi:10.1002/14356007.002_011. ISBN 9783527306732.

- G. J. Leigh (2004). The world's greatest fix: a history of nitrogen and agriculture. Oxford University Press US. pp. 134–139. ISBN 0-19-516582-9.

- Trevor Illtyd Williams; Thomas Kingston Derry (1982). A short history of twentieth-century technology c. 1900-c. 1950. Oxford University Press. pp. 134–135. ISBN 0-19-858159-9.

- Partington, J. R. (1960). A History of Greek Fire and Gunpowder (illustrated, reprint ed.). JHU Press. p. 335. ISBN 0801859549. Retrieved 2014-11-21.

- Richard E. Jones & Kristin H. López (2006). Human Reproductive Biology, Third Edition. Elsevier/Academic Press. p. 225. ISBN 0-12-088465-8.

- Section 1.9.2 (page 76) in: Jacki Bishop; Thomas, Briony (2007). Manual of Dietetic Practice. Wiley-Blackwell. ISBN 1-4051-3525-5.

- Nicolaou, Kyriacos Costa; Tamsyn Montagnon (2008). Molecules That Changed The World. Wiley-VCH. p. 11. ISBN 978-3-527-30983-2.

4

Understanding Soil Chemistry

The study of the chemical used in the soil is known as soil chemistry. It is affected by many factors; some of these factors are mineral composition, organic matter and the environmental factors. Soil test, soil pH and soil health are the important factors of soil chemistry. This chapter has been written to provide an easy understanding of the varied facets of soil chemistry.

Soil Chemistry

Soil chemistry is the study of the chemical characteristics of soil. Soil chemistry is affected by mineral composition, organic matter and environmental factors.

Soil chemistry is the study of the chemical characteristics of soil. Soil chemistry is affected by mineral composition, organic matter and environmental factors. Soil - (i) The unconsolidated mineral or organic material on the immediate surface of the Earth that serves as a natural medium for the growth of land plants. (ii) The unconsolidated mineral or organic matter on the surface of the Earth that has been subjected to and shows effects of genetic and environmental factors of: climate (including water and temperature effects), and macro- and micro organisms, conditioned by relief, acting on parent material over a period of time. A product-soil differs from the material from which it is derived in many physical, chemical,biological, and morphological properties and characteristics. This definition is from Soil Taxonomy, second edition.

Key facts you should know about Soil:

Soil makes up the outermost layer of our planet and is formed from rocks and decaying plants and Animals. Soil has varying amounts of organic matter (living and dead organisms), minerals, and nutrients. • An average soil sample is 45 percent minerals, 25 percent water, 25 percent air, and five percent organic matter. Different-sized mineral particles, such as sand, silt, and clay, give soil its texture. • Topsoil is the most productive soil layer. • Ten tonnes of topsoil spread evenly over a hectare is only as thick as a one Euro coin. • Natural processes can take more than 500 years to form 2 centimeters of topsoil. • In some cases, 5 tonnes of animal life can live in one hectare of soil. • Fungi and bacteria help break down organic matter in the soil. • Earthworms digest organic matter, recycle nutrients, and make the surface soil richer. • Roots loosen the soil, allowing oxygen to penetrate. This benefits animals living in the soil. They also hold soil together and help prevent erosion. • A fully functioning soil reduces the risk of floods and protects underground water supplies by neutralizing or filtering out potential pollutants and storing as much as 3750 tonnes of water per hectare. • Scientists have found

10,000 types of soil in Europe and about 70,000 types of soil in the United States. • Soil stores 10% of the world's carbon dioxide emissions.

Environmental Soil Chemistry

A knowledge of environmental soil chemistry is paramount to predicting the fate of contaminants, as well as the processes by which they are initially released into the soil. Once a chemical is exposed to the soil environment myriad chemical reactions can occur that may increase or decrease contaminant toxicity. These reactions include adsorption/desorption, precipitation, polymerization, dissolution, complexation and oxidation/reduction. These reactions are often disregarded by scientists and engineers involved with environmental remediation. Understanding these processes enable us to better predict the fate and toxicity of contaminants and provide the knowledge to develop scientifically correct, and cost-effective remediation strategies.

Concepts

- Anion and cation exchange capacity
- Soil pH
- Mineral formation and transformation processes
- Clay mineralogy
- Sorption and precipitation reactions in soil
- Oxidation-reduction reactions
- Chemistry of problem soils

Soil Test

Soil test may refer to one or more of a wide variety of soil analyses conducted for one of several possible reasons. Possibly the most widely conducted soil tests are those done to estimate the plant-available concentrations of plant nutrients, in order to determine fertilizer recommendations in agriculture. Other soil tests may be done for engineering (geotechnical), geochemical or ecological investigations.

Plant Nutrition

In agriculture, a soil test commonly refers to the analysis of a soil sample to determine nutrient content, composition, and other characteristics such as the acidity or pH level. A soil test can determine fertility, or the expected growth potential of the soil which indicates nutrient deficiencies, potential toxicities from excessive fertility and inhibitions from the presence of non-essential trace minerals. The test is used to mimic the function of roots to assimilate minerals. The expected rate of growth is modeled by the Law of the Maximum.

Labs, such as those at Iowa State and Colorado State University, recommend that a soil test contains 10-20 sample points for every 40 acres (160,000 m^2) of field. Tap water or chemicals can change the composition of the soil, and may need to be tested separately. As soil nutrients vary with depth and soil components change with time, the depth and timing of a sample may also affect results.

Composite sampling can be performed by combining soil from several locations prior to analysis. This is a common procedure, but should be used judiciously to avoid skewing results. This procedure must be done so that government sampling requirements are met. A reference map should be created to record the location and quantity of field samples in order to properly interpret test results.

Storage, Handling, and Moving

Soil chemistry changes over time, as biological and chemical processes break down or combine compounds over time. These processes change once the soil is removed from its natural ecosystem (flora and fauna that penetrate the sampled area) and environment (temperature, moisture, and solar light/radiation cycles). As a result, the chemical composition analysis accuracy can be improved if the soil is analysed soon after its extraction — usually within a relative time period of 24 hours. The chemical changes in the soil can be slowed during storage and transportation by freezing it. Air drying can also preserve the soil sample for many months.

Soil Testing

Soil testing is often performed by commercial labs that offer a variety of tests, targeting groups of compounds and minerals. The advantages associated with local lab is that they are familiar with the chemistry of the soil in the area where the sample was taken. This enables technicians to recommend the tests that are most likely to reveal useful information.

Soil testing in progress

Laboratory tests often check for plant nutrients in three categories:

- Major nutrients: nitrogen (N), phosphorus (P), and potassium (K)

- Secondary nutrients: sulfur, calcium, magnesium

- Minor nutrients: iron, manganese, copper, zinc, boron, molybdenum, chlorine

Do-it-yourself kits usually only test for the three "major nutrients", and for soil acidity or pH level. Do-it-yourself kits are often sold at farming cooperatives, university labs, private labs, and some hardware and gardening stores. Electrical meters that measure pH, water content, and sometimes nutrient content of the soil are also available at many hardware stores. Laboratory tests are more accurate than tests with do-it-yourself kits and electrical meters. Here is an example soil sample report from one laboratory.

Soil testing is used to facilitate fertilizer composition and dosage selection for land employed in both agricultural and horticultural industries.

Prepaid mail-in kits for soil and ground water testing are available to facilitate the packaging and delivery of samples to a laboratory. Similarly, in 2004, laboratories began providing fertilizer recommendations along with the soil composition report.

Lab tests are more accurate, though both types are useful. In addition, lab tests frequently include professional interpretation of results and recommendations. Always refer to all proviso statements included in a lab report as they may outline any anomalies, exceptions, and shortcomings in the sampling and/or analytical process/results.

Some laboratories analyze for all 13 mineral nutrients and a dozen non-essential, potentially toxic minerals utilizing the "universal soil extractant" (ammonium bicarbonate DTPA).

Soil Contaminants

Common mineral soil contaminants include arsenic, barium, cadmium, copper, mercury, lead, and zinc.

Lead is a particularly dangerous soil component. The following table from the University of Minnesota categorizes typical soil concentration levels and their associated health risks.

Six gardening practices to reduce the lead risk

1. Locate gardens away from old painted structures and heavily traveled roads

2. Give planting preferences to fruiting crops (tomatoes, squash, peas, sunflowers, corn, etc.)

3. Incorporate organic materials such as finished compost, humus, and peat moss

4. Lime soil as recommended by soil test (pH 6.5 minimizes lead availability)

5. Discard old and outer leaves before eating leafy vegetables; peel root crops; wash all produce

6. Keep dust to a minimum by maintaining a mulched and/or moist soil surface

Soil pH

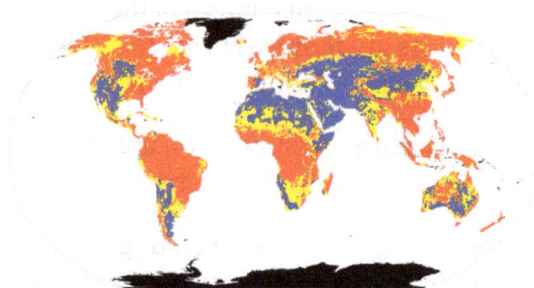

Global variation in soil pH. Red = acidic soil. Yellow = neutral soil. Blue = alkaline soil. Black = no data.

The soil pH is a measure of the acidity or alkalinity in soils. pH is defined as the negative logarithm (base 10) of the activity of hydronium ions (H^+or, more precisely, H_3O^+aq) in a solution. In water, it normally ranges from -1 to 14, with 7 being neutral. A pH below 7 is acidic and above 7 is alkaline. Soil pH is considered a master variable in soils as it controls many chemical processes that take place. It specifically affects plant nutrient availability by controlling the chemical forms of the nutrient. The optimum pH range for most plants is between 5.5 and 7.0, however many plants have adapted to thrive at pH values outside this range.

Classification of Soil pH Ranges

The United States Department of Agriculture Natural Resources Conservation Service, formerly Soil Conservation Service classifies soil pH ranges as follows:

Denomination	pH range
Ultra acidic	< 3.5
Extremely acidic	3.5–4.4
Very strongly acidic	4.5–5.0
Strongly acidic	5.1–5.5
Moderately acidic	5.6–6.0
Slightly acidic	6.1–6.5

Neutral	6.6–7.3
Slightly alkaline	7.4–7.8
Moderately alkaline	7.9–8.4
Strongly alkaline	8.5–9.0
Very strongly alkaline	> 9.0

Sources of Soil pH

Sources of Acidity

Acidity in soils comes from H^+ and Al^{3+} ions in the soil solution and sorbed to soil surfaces. While pH is the measure of H^+ in solution, Al^{3+} is important in acid soils because between pH 4 and 6, Al^{3+} reacts with water (H_2O) forming $AlOH^{2+}$, and $Al(OH)_2^+$, releasing extra H^+ ions. Every Al^{3+} ion can create 3 H^+ ions. Many other processes contribute to the formation of acid soils including rainfall, fertilizer use, plant root activity and the weathering of primary and secondary soil minerals. Acid soils can also be caused by pollutants such as acid rain and mine spoilings.

- Rainfall: Acid soils are most often found in areas of high rainfall. Excess rainfall leaches base cation from the soil, increasing the percentage of Al^{3+} and H^+ relative to other cations. Additionally, rainwater has a slightly acidic pH of 5.7 due to a reaction with CO_2 in the atmosphere that forms carbonic acid.

- Fertilizer use: Ammonium (NH_4^+) fertilizers react in the soil in a process called nitrification to form nitrate (NO_3^-), and in the process release H^+ ions.

- Plant root activity: Plants take up nutrients in the form of ions (NO_3^-, NH_4^+, Ca^{2+}, $H_2PO_4^-$, etc.), and often, they take up more cations than anions. However plants must maintain a neutral charge in their roots. In order to compensate for the extra positive charge, they will release H^+ ions from the root. Some plants will also exude organic acids into the soil to acidify the zone around their roots to help solubilize metal nutrients that are insoluble at neutral pH, such as iron (Fe).

- Weathering of minerals: Both primary and secondary minerals that compose soil contain Al. As these minerals weather, some components such as Mg, Ca, and K, are taken up by plants, others such as Si are leached from the soil, but due to chemical properties, Fe and Al remain in the soil profile. Highly weathered soils are often characterized by having high concentrations of Fe and Al oxides.

- Acid rain: When atmospheric water reacts with sulfur and nitrogen compounds that result from industrial processes, the result can be the formation of sulfuric and nitric acid in rainwater. However the amount of acidity that is deposited in rainwater is much less, on average, than that created through agricultural activities.

- Mine spoil: Severely acidic conditions can form in soils near mine spoils due to the oxidation of pyrite.

- Potential acid sulfate soils naturally formed in waterlogged coastal and estuarine environments can become highly acidic when drained or excavated.

- Decomposition of organic matter by microorganisms releases CO_2 which when mixed with soil water can form carbonic acid (H_2CO_3).

Sources of Alkalinity

Alkaline soils have a high saturation of base cations (K^+, Ca^{2+}, Mg^{2+} and Na^+). This is due to an accumulation of soluble salts which are classified as either saline soil, sodic soil, saline-sodic soil or alkaline soil. All saline and sodic soils have high salt concentrations, with saline soils being dominated by calcium and magnesium salts and sodic soils being dominated by sodium. Alkaline soils are characterized by the presence of carbonates. Soil in areas with limestone near the surface are alkaline from the calcium carbonate in limestone constantly mixing with the soil. Groundwater sources in these areas contain dissolved limestone.

Effect of Soil pH on Plant Growth

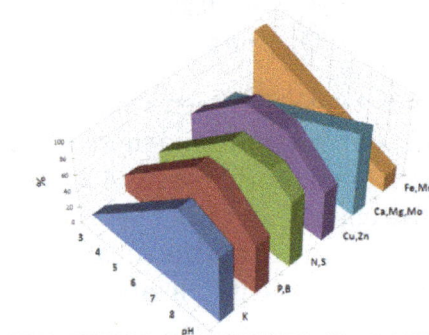

Nutrient availability in relation to soil pH

Acid Affected Soils

Plants grown in acid soils can experience a variety of symptoms including aluminium (Al), hydrogen (H), and/or manganese (Mn) toxicity, as well as nutrient deficiencies of calcium (Ca) and magnesium (Mg).

Aluminium toxicity is the most widespread problem in acid soils. Aluminium is present in all soils, but dissolved Al^{3+} is toxic to plants; Al^{3+} is most soluble at low pH, above pH 5.2 little Al is in soluble form in most soils. Aluminium is not a plant nutrient, and as such, is not actively taken up by the plants, but enters plant roots passively through osmosis. Aluminium inhibits root growth; lateral roots and root tips become thickened and roots lack fine branching; root tips may turn brown. In the root, Al has been shown to interfere with many physiological processes including the uptake and transport of calcium and other essential nutrients, cell division, cell wall formation, and enzyme activity.

Below pH 4, H^+ ions themselves damage root cell membranes.

In soils with high content of manganese-containing minerals, Mn toxicity can become a problem at pH 5.6 and lower. Manganese, like aluminium, becomes increasingly soluble as pH drops, and Mn toxicity symptoms can be seen at pH levels below 5.6. Manganese is an essential plant nutrient, so plants transport Mn into leaves. Classic symptoms of Mn toxicity are crinkling or cupping of leaves.

Nutrient Availability in Relation to Soil pH

Nutrients needed in large amounts by plants are referred to as macronutrients and include nitrogen (N), phosphorus (P), potassium (K), calcium (Ca), magnesium (Mg) and sulfur (S). Elements that plants need in trace amounts are called trace nutrients or micronutrients. Trace nutrients are not major components of plant tissue but are essential for growth. They include iron (Fe), manganese (Mn), zinc (Zn), copper (Cu), cobalt (Co), molybdenum (Mo), and boron (B). Both macronutrient and micronutrient availability are affected by soil pH. In slightly to moderately alkaline soils, molybdenum and macronutrient (except for phosphorus) availability is increased, but P, Fe, Mn, Zn Cu, and Co levels are reduced and may adversely affect plant growth. In acidic soils, micronutrient availability (except for Mo and Bo) is increased. Nitrogen is supplied as ammonium (NH4) or nitrate (NO3) by nitrogen fixation or fertilizer amendments, and dissolved N will have the highest concentrations in soil with pH 6–8. Concentrations of available N are less sensitive to pH than concentration of available P. In order for P to be available for plants, soil pH needs to be in the range 6.0 and 7.5. If pH is lower than 6, P starts forming insoluble compounds with iron (Fe) and aluminium (Al) and if pH is higher than 7.5 P starts forming insoluble compounds with calcium (Ca). Most nutrient deficiencies can be avoided between a pH range of 5.5 to 6.5, provided that soil minerals and organic matter contain the essential nutrients to begin with.

Water Availability in Relation to Soil pH

Determining pH

Methods of determining pH include:

- Observation of soil profile: Certain profile characteristics can be indicators of either acid, saline, or sodic conditions. Strongly acidic soils often have poor incorporation of the organic surface layer with the underlying mineral layer. The mineral horizons are distinctively layered in many cases, with a pale eluvial (E) horizon beneath the organic surface; this E is underlain by a darker B horizon in a classic podzol horizon sequence. This is a very rough gauge of acidity as there is no correlation between thickness of the E and soil pH. E horizons a few feet thick in Florida usually have pH just above 5 (merely "strongly acid") while E horizons a few inches thick in New England are "extremely acid" with pH readings of 4.5 or below. In the southern Blue Ridge Mountains there are "ultra acid" soils, pH below 3.5, which have no E horizon. Presence of a caliche layer indicates the presence of calcium carbonates, which are present in alkaline conditions. Also, columnar structure can be an indicator of sodic condition.

- Observation of predominant flora. Calcifuge plants (those that prefer an acidic soil) include *Erica*, *Rhododendron* and nearly all other Ericaceae species, many birch (*Betula*), foxglove (*Digitalis*), gorse (*Ulex* spp.), and Scots Pine (*Pinus sylvestris*). Calcicole (lime loving) plants include ash trees (*Fraxinus* spp.), honeysuckle (*Lonicera*), *Buddleja*, dogwoods (*Cornus* spp.), lilac (*Syringa*) and *Clematis* species.

- Use of an inexpensive pH testing kit, where in a small sample of soil is mixed with indicator solution which changes colour according to the acidity/alkalinity.

- Use of litmus paper. A small sample of soil is mixed with distilled water, into which a strip of litmus paper is inserted. If the soil is acidic the paper turns red, if alkaline, blue.

- Use of a commercially available electronic pH meter, in which a rod is inserted into moistened soil and measures the concentration of hydrogen ions.

Examples of Plant pH Preferences

- pH 4.5–5.0: Ericaceae (azalea, bilberry, blueberry, cranberry; heather), hydrangea for blue, (less acidic for pink), liquidambar or sweet gum, orchid, pin oak

- pH 5.0–5.5: Boronia, daphne; Ericaceae: (camellia, heather, rhododendron), ferns, iris, orchids, parsley, conifers (e.g., pine); Poaceae: (maize, millet, rye, oat), radish; Solanales: (potato, sweet potato); Bromeliaceae (pineapple).

- pH 5.5–6.0: Asteraceae: (aster, endive); Brassicaceae: (brussels sprout, kohlrabi), carrot; Cucurbitales: (begonia, chayote or choko); Fabaceae: (bean, crimson clover, peanut, soybean), petunia, rhubarb, violet, most bulbs (canna, daffodil, jonquil), larkspur, primrose

- pH 6.0–6.5 antirrhinum or snapdragon, Brassicaceae: (broccoli, cabbage, candytuft, cauliflower, turnip, wallflower); Cucurbitaceae: (cucumber, pumpkin, squash); Fabaceae: (pea, red clover, white clover), gladiolus, Iceland poppy; Rosales: (cannabis, rose, strawberry); Solanaceae: (eggplant or aubergine, tomato), sweet corn; Violaceae: (pansy, viola), zinnia or zinnea

- pH 6.5–7.0: Amaranthaceae: (beet, spinach); Apiaceae: (celery, parsnip); Asparagales: (asparagus, onion); Asteraceae: (chrysanthemum, dahlia, lettuce), carnation; Fabaceae: (alfalfa, sweet pea), melon, stock, tulip

- pH 7.1–8.0 lilac

Changing Soil pH

Increasing pH of Acidic Soil

The most common amendment to increase soil pH is lime ($CaCO_3$ or $MgCO_3$), usually in the form of finely ground agricultural lime. The amount of lime needed to change pH is determined by the mesh size of the lime (how finely it is ground)and the buffering capacity of the soil. A high mesh size (60–100) indicates a finely ground lime, that will react quickly with soil acidity. Buffering capacity of soils is a function of a soils cation exchange capacity, which is in turn determined by the clay content of the soil, the type of clay and the amount of organic matter present. Soils with high clay content, particularly shrink–swell clay, will have a higher buffering capacity than soils with little clay. Soils with high organic matter will also have a higher buffering capacity than those with low organic matter. Soils with high buffering capacity require a greater amount of lime to be added than a soil with a lower buffering capacity for the same incremental change in pH. Other amendments that can be used to increase the pH of soil include wood ash, industrial CaO (burnt lime), and oyster shells. White firewood ash includes metal salts which are important for processes requiring ions such as Na^+ (sodium), K^+ (potassium), Ca^{2+} (calcium), which may or may not be good for the select flora, but decreases the acidic quality of soil. These products increase the pH of soils through the reaction of CO_3^{2-} with H^+ to produce CO_2 and H_2O. Calcium silicate neutralizes active acidity in the soil by removing free hydrogen ions, thereby increasing pH. As its silicate anion captures H^+ ions (raising the pH), it forms monosilicic acid (H_4SiO_4), a neutral solute.

Decreasing pH of Alkaline Soil

- Iron sulfates or aluminium sulfate as well as elemental sulfur (S) reduce pH through the formation of sulfuric acid.

- Urea, urea phosphate, ammonium nitrate, ammonium phosphates, ammonium sulfate and monopotassium phosphate fertilizers.

- organic matter in the form of plant litter, compost, and manure will decrease soil pH through the decomposition process. Certain acid organic matter such as pine needles, pine sawdust and acid peat are effective at reducing pH.

Illinois Soil Nitrogen Test

The Illinois Soil Nitrogen Test ("ISNT") is a method for measuring the amount of Nitrogen in soil that is available for use by plants as a nutrient. The test predicts whether the addition of nitrogen fertilizer to agricultural land will result in increased crop yields.

Nitrogen is essential for plant development. Indeed, for crops that are destined to be food for farm animal or human consumption, incorporation of nitrogen into the crop is an important goal, since this forms the basis for protein in the human diet.

Nitrogen is commonly present in soils in many forms, and there are many ways to measure this nitrogen. None of these are completely satisfactory as a measure of the nitrogen that is available for use by crops. The ISNT is a new (2007) method for measuring nitrogen available for plant uptake.

ISNT estimates the amount of nitrogen present in the soil as amino sugar nitrogen. With respect to corn and soybeans, the optimal range for plant growth appears to be around 225 to 240 mg/Kg. Some form of nitrogen fertilizer is needed if levels are below this range. On the other hand, if levels are above this range, addition of nitrogen fertilizer will not increase crop yield.

In the corn belt, since about 1975, the predominant method of estimating the amount of nitrogen needed for corn has been the "yield-based" method. A farmer first estimates the yield of corn he intends to produce. He then applies 1.1 to 1.4 lbs of nitrogen per bushel of expected yield. ISNT represents an alternative approach to managing nitrogen application. However, ISNT does not offer a simple answer as to the amount of nitrogen fertilizer that is needed, or as to the optimal form of that fertilizer.

In field trials in Illinois, some fields have been found to be under-fertilized when managed according to the "yield-based" method, as judged by the ISNT. In the majority of trials, however, the yield-based method calls for the addition of nitrogen far in excess of the levels needed for optimal crop production. This nitrogen, which is applied by farmers at great cost, does not find its way into the crop, but is lost to the atmosphere or leaches into waterways.

Within the corn belt, stalks and other crop residues are left in the field with the intention of enhancing the amount of organic material in the soil. Excessive nitrogen application, however, appears to promote the rapid decomposition of organic matter in the soil, resulting in release of carbon dioxide. As a result, the amount of organic material in soils managed according to the yield-based method in the corn belt appears to be decreasing in spite of the large amounts of crop residues left in the fields.

Soil Health

Soil health is a state of a soil meeting its range of ecosystem functions as appropriate to its environment. Soil health testing is an assessment of this status. Soil health depends on soil biodiversity (with a robust soil biota), and it can be improved via soil conditioning (soil amendment).

Aspects

The term soil health is used to describe the state of a soil in:

- Sustaining plant and animal productivity and biodiversity (Soil biodiversity);

- Maintaining or enhance water and air quality;

- Supporting human health and habitation.

Soil Health has partly if not largely replaced the expression "Soil Quality" that was extant in the 1990s. The primary difference between the two expressions is that soil quality was focused on individual traits within a functional group, as in "quality of soil for maize production" or "quality of soil for roadbed preparation" and so on. The addition of the word "health" shifted the perception to be integrative, holistic and systematic. The two expressions still overlap considerably.

The underlying principle in the use of the term "soil health" is that soil is not just an inert, lifeless growing medium, which modern farming tends to represent, rather it is a living, dynamic and ever-so-subtly changing whole environment. It turns out that soils highly fertile from the point of view of crop productivity are also lively from a biological point of view. It is now commonly recognized that soil microbial biomass is large: in temperate grassland soil the bacterial and fungal biomass have been documented to be 1–2 t (2.0 long tons; 2.2 short tons)/hectare and 2–5 t (4.9 long tons; 5.5 short tons)/ha, respectively. Some microbiologists now believe that 80% of soil nutrient functions are essentially controlled by microbes. If this is consistently true, than the prevailing Liebig nutrient theory model, which excludes biology, is perhaps dangerously incorrect for managing soil fertility sustainably for the future.

Using the human health analogy, a healthy soil can be categorized as one:

- In a state of composite well-being in terms of biological, chemical and physical properties;

- Not diseased or infirmed (i.e. not degraded, nor degrading), nor causing negative off-site impacts;

- With each of its qualities cooperatively functioning such that the soil reaches its full potential and resists degradation;

- Providing a full range of functions (especially nutrient, carbon and water cycling) and in such a way that it maintains this capacity into the future.

Conceptualisation

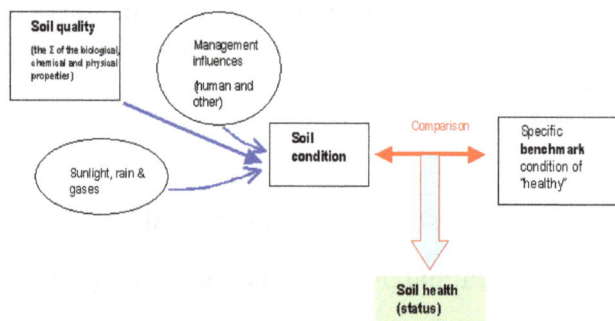

Soil health is the condition of the soil in a defined space and at a defined scale relative to a set of benchmarks that encompass healthy functioning. It would not be appropriate to refer to soil health for soil-roadbed preparation, as in the analogy of soil quality in a functional class. The definition of soil health may vary between users of the term as alternative users may place differing priorities upon the multiple functions of a soil. Therefore, the term soil health can only be understood within the context of the user of the term, and their aspirations of a soil, as well as by the boundary definition of the soil at issue.

Interpretation

Different soils will have different benchmarks of health depending on the "inherited" qualities, and on the geographic circumstance of the soil. The generic aspects defining a healthy soil can be considered as follows:

- "Productive" options are broad;

- Life diversity is broad;

- Absorbency, storing, recycling and processing is high in relation to limits set by climate;

- Water runoff quality is of high standard;

- Low entropy; and,

- No damage to, or loss of the fundamental components.

This translates to:

- A comprehensive cover of vegetation;

- Carbon levels relatively close to the limits set by soil type and climate;

- Little leakage of nutrients from the ecosystem;

- Biological and agricultural productivity relatively close to the limits set by the soil environment and climate;

- Only geological rates of erosion;

- No accumulation of contaminants; and,

- The ecosystem does not rely excessively on inputs of fossil energy

An unhealthy soil thus is the simple converse of the above.

Measurement

On the basis of the above, soil health will be measured in terms of individual ecosystem services provided relative to the benchmark. Specific benchmarks used to evaluate soil health include CO_2 release, humus levels, microbial activity, and available calcium.

Soil health testing is spreading in the USA, Australia and S. Africa. Cornell University has had a Soil Health Test since 2006 and offers many types of individual soil health tests as well as three different soil health testing packages. The approach is to add into common chemical nutrient testing a biological set of factors not normally included in routine soil testing. The best example is adding biological soil respiration (CO2-Burst) as a test procedure; this has already been adapted to modern commercial labs in the past 10 years.

There is resistance among soil testing labs and university scientists to adding new biological tests, primarily since interpretation of soil fertility is based on models from "crop response" studies which match yield to test levels of specific chemical nutrients. These soil test methods have evolved slowly over the past 40 years and are backed by considerable effort. However, in this same time USA soils have also lost up to 75% of their carbon (humus), causing biological fertility to drop drastically. Many critics of the current system say this is sufficient evidence that the old soil testing models have failed us, and need to be replaced with new approaches. These older models have stressed "maximum yield" and "yield calibration" to such an extent that related factors have been overlooked. Thus, surface and groundwater pollution with excess nutrients (nitrates and phosphates) has grown enormously, and is reported presently (in the USA) to be the worst it has been since the 1970s, before the advent of environmental consciousness.

References

- Carl J. Rosen. "Lead in the Home Garden and Urban Soil Environment". Extension.umn.edu. Retrieved 2012-11-08.

- Soil Survey Division Staff. "Soil survey manual.1993. Chapter 3, selected chemical properties.". Soil Conservation Service. U.S. Department of Agriculture Handbook 18. Retrieved 2011-03-12.

- Hansson et al (2011) Differences in soil properties in adjacent stands of Scots pine, Norway spruce and silver birch in SW Sweden. Forest Ecology and Management 262 522–530

Plant Nutrition and its Management

The nutrition that every plant necessarily needs can be supplied externally also. Nutrient management connects soil, crop and weather to irrigation and soil and water. It is the management of matching the right soil with the right climate and crop management. The section serves as a source to understand the major aspects of plant nutrition.

Plant Nutrition

Plant nutrition is the study of the chemical elements and compounds necessary for plant growth, plant metabolism and their external supply. In 1972, E. Epstein defined two criteria for an element to be essential for plant growth:

1. in its absence the plant is unable to complete a normal life cycle.

2. or that the element is part of some essential plant constituent or metabolite.

This is in accordance with Justus von Liebig's law of the minimum. The essential plant nutrients include carbon and oxygen which are absorbed from the air, whereas other nutrients including hydrogen are typically obtained from the soil (exceptions include some parasitic or carnivorous plants).

Plants must obtain the following mineral nutrients from their growing medium:

*   the macronutrients: nitrogen (N), phosphorus (P), potassium (K), calcium (Ca), sulfur (S), magnesium (Mg); and

*   the micronutrients (or trace minerals): boron (B), chlorine (Cl), manganese (Mn), iron (Fe), zinc (Zn), copper (Cu), molybdenum (Mo), nickel (Ni).

The macronutrients are consumed in larger quantities and are usually present in plant tissue in concentrations of between 0.2% and 4.0% (on a dry matter weight basis). Micronutrients are present in plant tissue in quantities measured in parts per million, ranging from 0.1 to 200 ppm, or less than 0.02% dry weight.

Most soil conditions across the world can provide plants adapted to that climate and soil with sufficient nutrition for a complete life cycle, without the addition of nutrients as fertilizer. However, if the soil is cropped it is necessary to artificially modify soil fertility through the addition of fertilizer to promote vigorous growth and increase or sustain yield. This is done because, even with adequate water and light, nutrient deficiency can limit growth and crop yield.

Farmer spreading decomposing manure to improve soil fertility and plant nutrition

Processes

Plants take up essential elements from the soil through their roots and from the air (mainly consisting of nitrogen and oxygen) through their leaves. Nutrient uptake in the soil is achieved by cation exchange, wherein root hairs pump hydrogen ions (H^+) into the soil through proton pumps. These hydrogen ions displace cations attached to negatively charged soil particles so that the cations are available for uptake by the root. In the leaves, stomata open to take in carbon dioxide and expel oxygen. The carbon dioxide molecules are used as the carbon source in photosynthesis.

The root, especially the root hair, is the essential organ for the uptake of nutrients. The structure and architecture of the root can alter the rate of nutrient uptake. Nutrient ions are transported to the center of the root, the stele in order for the nutrients to reach the conducting tissues, xylem and phloem. The Casparian strip, a cell wall outside the stele but within the root, prevents passive flow of water and nutrients, helping to regulate the uptake of nutrients and water. Xylem moves water and inorganic molecules within the plant and phloem accounts for organic molecule transportation. Water potential plays a key role in a plant's nutrient uptake. If the water potential is more negative within the plant than the surrounding soils, the nutrients will move from the region of higher solute concentration—in the soil—to the area of lower solute concentration - in the plant.

There are three fundamental ways plants uptake nutrients through the root:

1. Simple diffusion occurs when a nonpolar molecule, such as O_2, CO_2, and NH_3 follows a concentration gradient, moving passively through the cell lipid bilayer membrane without the use of transport proteins.

2. Facilitated diffusion is the rapid movement of solutes or ions following a concentration gradient, facilitated by transport proteins.

3. Active transport is the uptake by cells of ions or molecules against a

concentration gradient; this requires an energy source, usually ATP, to power molecular pumps that move the ions or molecules through the membrane.

Nutrients can be moved within plants to where they are most needed. For example, a plant will try to supply more nutrients to its younger leaves than to its older ones. When nutrients are mobile within the plant, symptoms of any deficiency become apparent first on the older leaves. However, not all nutrients are equally mobile. Nitrogen, phosphorus, and potassium are mobile nutrients while the others have varying degrees of mobility. When a less-mobile nutrient is deficient, the younger leaves suffer because the nutrient does not move up to them but stays in the older leaves. This phenomenon is helpful in determining which nutrients a plant may be lacking.

Many plants engage in symbiosis with microorganisms. Two important types of these relationship are

1. with bacteria such as rhizobia, that carry out biological nitrogen fixation, in which atmospheric nitrogen (N_2) is converted into ammonium ($NH+4$); and

2. with mycorrhizal fungi, which through their association with the plant roots help to create a larger effective root surface area. Both of these mutualistic relationships enhance nutrient uptake.

Though nitrogen is plentiful in the Earth's atmosphere, relatively few plants harbour nitrogen-fixing bacteria, so most plants rely on nitrogen compounds present in the soil to support their growth. These can be supplied by mineralization of soil organic matter or added plant residues, nitrogen fixing bacteria, animal waste, through the breaking of triple bonded Nitrogen molecules by lightening strikes or through the application of fertilizers.

Functions of Nutrients

At least 17 elements are known to be essential nutrients for plants. In relatively large amounts, the soil supplies nitrogen, phosphorus, potassium, calcium, magnesium, and sulfur; these are often called the macronutrients. In relatively small amounts, the soil supplies iron, manganese, boron, molybdenum, copper, zinc, chlorine, and cobalt, the so-called micronutrients. Nutrients must be available not only in sufficient amounts but also in appropriate ratios.

Plant nutrition is a difficult subject to understand completely, partially because of the variation between different plants and even between different species or individuals of a given clone. Elements present at low levels may cause deficiency symptoms, and toxicity is possible at levels that are too high. Furthermore, deficiency of one element may present as symptoms of toxicity from another element, and vice versa. An abundance of one nutrient may cause a deficiency of another nutrient. For example, K^+ uptake can be influenced by the amount of $NH+4$ available.

Although nitrogen is plentiful in the Earth's atmosphere, relatively few plants engage in nitrogen fixation (conversion of atmospheric nitrogen to a biologically useful form). Most plants, therefore, require nitrogen compounds to be present in the soil in which they grow.

Carbon and oxygen are absorbed from the air while other nutrients are absorbed from the soil. Green plants obtain their carbohydrate supply from the carbon dioxide in the air by the process of photosynthesis. Each of these nutrients is used in a different place for a different essential function.

Macronutrients (Derived from Air and Water)

Carbon

Carbon forms the backbone of most plant biomolecules, including proteins, starches and cellulose. Carbon is fixed through photosynthesis; this converts carbon dioxide from the air into carbohydrates which are used to store and transport energy within the plant.

Hydrogen

Hydrogen also is necessary for building sugars and building the plant. It is obtained almost entirely from water. Hydrogen ions are imperative for a proton gradient to help drive the electron transport chain in photosynthesis and for respiration.

Oxygen

Oxygen is a component of many organic and inorganic molecules within the plant, and is acquired in many forms. These include: O_2 and CO_2 (mainly from the air via leaves) and H_2O, $NO-3$, H_2PO-4 and $SO2-4$ (mainly from the soil water via roots). Plants produce oxygen gas (O_2) along with glucose during photosynthesis but then require O_2 to undergo aerobic cellular respiration and break down this glucose to produce ATP.

Macronutrients (Primary)

Nitrogen

Nitrogen is a major constituent of several of the most important plant substances. For example, nitrogen compounds comprise 40% to 50% of the dry matter of protoplasm, and it is a constituent of amino acids, the building blocks of proteins. It is also an essential constituent of chlorophyll. Nitrogen deficiency most often results in stunted growth, slow growth, and chlorosis. Nitrogen deficient plants will also exhibit a purple appearance on the stems, petioles and underside of leaves from an accumulation of anthocyanin pigments. Most of the nitrogen taken up by plants is from the soil in the forms of $NO-3$, although in acid environments such as boreal forests where nitrification is less likely to occur, ammonium $NH+4$ is more likely to be the dominating source of nitrogen. Amino acids and proteins can only be built from $NH+4$, so $NO-3$ must be reduced. In many agricultural settings, nitrogen is the limiting nutrient for rapid growth. Nitrogen is transported via the xylem from the roots to the leaf canopy as nitrate ions, or in an organic form, such as amino acids or amides. Nitrogen can also be transported in the phloem sap as amides, amino acids and ureides; it is therefore mobile within the plant, and the older leaves exhibit chlorosis and necrosis earlier than the younger leaves.

There is an abundant supply of nitrogen in the earth's atmosphere — N_2 gas comprises nearly 79% of air. However, N_2 is unavailable for use by most organisms because there is a triple bond between the two nitrogen atoms, making the molecule almost inert. In order for nitrogen to be used for growth it must be "fixed" (combined) in the form of ammonium (NH_4) or nitrate (NO_3) ions. The weathering of rocks releases these ions so slowly that it has a negligible effect on the availability of fixed nitrogen. Therefore, nitrogen is often the limiting factor for growth and biomass production in all environments where there is a suitable climate and availability of water to support life.

Nitrogen enters the plant largely through the roots. A "pool" of soluble nitrogen accumulates. Its composition within a species varies widely depending on several factors, including day length, time of day, night temperatures, nutrient deficiencies, and nutrient imbalance. Short day length promotes asparagine formation, whereas glutamine is produced under long day regimes. Darkness favors protein breakdown accompanied by high asparagine accumulation. Night temperature modifies the effects due to night length, and soluble nitrogen tends to accumulate owing to retarded synthesis and breakdown of proteins. Low night temperature conserves glutamine; high night temperature increases accumulation of asparagine because of breakdown. Deficiency of K accentuates differences between long- and short-day plants. The pool of soluble nitrogen is much smaller than in well-nourished plants when N and P are deficient since uptake of nitrate and further reduction and conversion of N to organic forms is restricted more than is protein synthesis. Deficiencies of Ca, K, and S affect the conversion of organic N to protein more than uptake and reduction. The size of the pool of soluble N is no guide *per se* to growth rate, but the size of the pool in relation to total N might be a useful ratio in this regard. Nitrogen availability in the rooting medium also affects the size and structure of tracheids formed in the long lateral roots of white spruce (Krasowski and Owens 1999).

Microorganisms have a central role in almost all aspects of nitrogen availability, and therefore for life support on earth. Some bacteria can convert N_2 into ammonia by the process termed *nitrogen fixation*; these bacteria are either free-living or form symbiotic associations with plants or other organisms (e.g., termites, protozoa), while other bacteria bring about transformations of ammonia to nitrate, and of nitrate to N_2 or other nitrogen gases. Many bacteria and fungi degrade organic matter, releasing fixed nitrogen for reuse by other organisms. All these processes contribute to the nitrogen cycle.

Phosphorus

Like nitrogen, phosphorus is involved with many vital plant processes. Within a plant, it is present mainly as a structural component of the nucleic acids, deoxyribonucleic nucleic acid (DNA) and ribose nucleic acid (RNA), and as a constituent of fatty phospholipids, of importance in membrane development and function. It is present in both organic and inorganic forms, both of which are readily translocated within the plant. All energy transfers in the cell are critically dependent on phosphorus. As with all living things, phosphorus is part of the Adenosine triphosphate (ATP), which is of immediate use in all processes that require energy with the cells. Phosphorus can also be used to modify the activity of various enzymes by phosphorylation, and is used for cell signaling. Phosphorus is concentrated at the most actively growing points of a plant and stored within seeds in anticipation of their germination. Phosphorus is most commonly found in the soil in the form of polyprotic phosphoric acid (H_3PO_4), but is taken up most readily in the form of H_2PO-4. Phosphorus is available to plants in limited quantities in most soils because it is released very slowly from insoluble phosphates and is rapidly fixed once again. Under most environmental conditions it is the element that limits growth because of this constriction and due to its high demand by plants and microorganisms. Plants can increase phosphorus uptake by a mutualism with mycorrhiza. A Phosphorus deficiency in plants is characterized by an intense green coloration or reddening in leaves due to lack of chlorophyll. If the plant is experiencing high phosphorus deficiencies the leaves may become denatured and show signs of death. Occasionally the leaves may appear purple from an accumulation of anthocyanin. Because phosphorus is a mobile nutrient, older leaves will show the first signs of deficiency.

On some soils, the phosphorus nutrition of some conifers, including the spruces, depends on the ability of mycorrhizae to take up, and make soil phosphorus available to the tree, hitherto unobtainable to the non-mycorrhizal root. Seedling white spruce, greenhouse-grown in sand testing negative for phosphorus, were very small and purple for many months until spontaneous mycorrhizal inoculation, the effect of which was manifested by a greening of foliage and the development of vigorous shoot growth.

Phosphorus deficiency can produce symptoms similar to those of nitrogen deficiency, but as noted by Russel: "Phosphate deficiency differs from nitrogen deficiency in being extremely difficult to diagnose, and crops can be suffering from extreme starvation without there being any obvious signs that lack of phosphate is the cause". Russell's observation applies to at least some coniferous seedlings, but Benzian found that although response to phosphorus in very acid forest tree nurseries in England was consistently high, no species (including Sitka spruce) showed any visible symptom of deficiency other than a slight lack of lustre. Phosphorus levels have to be exceedingly low before visible symptoms appear in such seedlings. In sand culture at 0 ppm phosphorus, white spruce seedlings were very small and tinted deep purple; at 0.62 ppm, only the smallest seedlings were deep purple; at 6.2 ppm, the seedlings were of good size and color.

It is useful to apply a high phosphorus content fertilizer, such as bone meal, to perennials to help with successful root formation.

Potassium

Unlike other major elements, potassium does not enter into the composition of any of the important plant constituents involved in metabolism, but it does occur in all parts of plants in substantial amounts. It seems to be of particular importance in leaves and at growing points. Potassium is outstanding among the nutrient elements for its mobility and solubility within plant tissues. Processes involving potassium include the formation of carbohydrates and proteins, the regulation of internal plant moisture, as a catalyst and condensing agent of complex substances, as an accelerator of enzyme action, and as contributor to photosynthesis, especially under low light intensity.

When soil-potassium levels are high, plants take up more potassium than needed for healthy growth. The term *luxury consumption* has been applied to this. When potassium is moderately deficient, the effects first appear in the older tissues, and from there progress towards the growing points. Acute deficiency severely affects growing points, and die-back commonly occurs. Symptoms of potassium deficiency in white spruce include: browning and death of needles (chlorosis); reduced growth in height and diameter; impaired retention of needles; and reduced needle length. A relationship between potassium nutrition and cold resistance has been found in several tree species, including two species of spruce.

Potassium regulates the opening and closing of the stomata by a potassium ion pump. Since stomata are important in water regulation, potassium regulates water loss from the leaves and increases drought tolerance. Potassium deficiency may cause necrosis or interveinal chlorosis. The potassium ion (K^+) is highly mobile and can aid in balancing the anion (negative) charges within the plant. Potassium helps in fruit coloration, shape and also increases its brix. Hence, quality fruits are produced in potassium-rich soils. Potassium serves as an activator of enzymes used in photosynthesis and respiration. Potassium is used to build cellulose and aids in photosynthesis by the formation of a chlorophyll precursor. Potassium deficiency may result in higher risk of pathogens, wilting, chlorosis, brown spotting, and higher chances of damage from frost and heat.

Macronutrients (Secondary and Tertiary)

Sulphur

Sulphur is a structural component of some amino acids and vitamins, and is essential in the manufacturing of chloroplasts. Sulphur is also found in the iron-sulphur complexes of the electron transport chains in photosynthesis. It is immobile and deficiency, therefore, affects younger tissues first. Symptoms of deficiency include yellowing of leaves and stunted growth.

Calcium

Calcium regulates transport of other nutrients into the plant and is also involved in the activation of certain plant enzymes. Calcium deficiency results in stunting. This nutrient is involved in photosynthesis and plant structure. Blossom end rot is also a result of inadequate calcium.

Calcium in plants occurs chiefly in the leaves, with lower concentrations in seeds, fruits, and roots. A major function is as a constituent of cell walls. When coupled with certain acidic compounds of the jelly-like pectins of the middle lamella, calcium forms an insoluble salt. It is also intimately involved in meristems, and is particularly important in root development, with roles in cell division, cell elongation, and the detoxification of hydrogen ions. Other functions attributed to calcium are; the neutralization of organic acids; inhibition of some potassium-activated ions; and a role in nitrogen absorption. A notable feature of calcium-deficient plants is a defective root system. Roots are usually affected before above-ground parts.

Magnesium

The outstanding role of magnesium in plant nutrition is as a constituent of the chlorophyll molecule. As a carrier, it is also involved in numerous enzyme reactions as an effective activator, in which it is closely associated with energy-supplying phosphorus compounds. Magnesium is very mobile in plants, and, like potassium, when deficient is translocated from older to younger tissues, so that signs of deficiency appear first on the oldest tissues and then spread progressively to younger tissues.

Micro-nutrients

Some elements are directly involved in plant metabolism (Arnon and Stout, 1939). However, this principle does not account for the so-called beneficial elements, whose presence, while not required, has clear positive effects on plant growth. Mineral elements that either stimulate growth but are not essential, or that are essential only for certain plant species, or under given conditions, are usually defined as beneficial elements.

Plants are able sufficiently to accumulate most trace elements. Some plants are sensitive indicators of the chemical environment in which they grow (Dunn 1991), and some plants have barrier mechanisms that exclude or limit the uptake of a particular element or ion species, e.g., alder twigs commonly accumulate molybdenum but not arsenic, whereas the reverse is true of spruce bark (Dunn 1991). Otherwise, a plant can integrate the geochemical signature of the soil mass permeated by its root system together with the contained groundwaters. Sampling is facilitated by the tendency of many elements to accumulate in tissues at the plant's extremities.

Iron

Iron is necessary for photosynthesis and is present as an enzyme cofactor in plants. Iron deficiency can result in interveinal chlorosis and necrosis. Iron is not a structural part of chlorophyll but very much essential for its synthesis. Copper deficiency can be responsible for promoting an iron deficiency.

Molybdenum

Molybdenum is a cofactor to enzymes important in building amino acids and is involved in nitrogen metabolism. Molybdenum is part of the nitrate reductase enzyme (needed for the reduction of nitrate) and the nitrogenase enzyme (required for biological nitrogen fixation).

Boron

Boron is found in the highly insoluble mineral, tourmaline. It is absorbed by plants in the form of the anion BO_3-3. It is available to plants in moderately soluble mineral forms of Ca, Mg and Na borates and the highly soluble form of organic compounds. Concentration in soil must, in general, be below 5 ppm in the soil water solution, above that toxicity results. Its availability in soils ranges from 20 to 200 pounds per acre in the first eight inches, of which less than 5% is available. It is available to plants over a range of pH, from 5.0 to 7.5. It is mobile in the soil, hence, it is prone to leaching. Leaching removes substantial amounts of boron in sandy soil, but little in fine silt or clay soil. Boron's fixation to those minerals at high pH can render boron unavailable, while low pH frees the fixed boron, leaving it prone to leaching in wet climates. It precipitates with other minerals in the form of borax in which form it was first used over 400 years ago as a soil supplement. Decomposition of organic material causes boron to be deposited in the topmost soil layer; organic forms of boron are more soluble than their mineral form, hence are more available in the top few inches. When soil dries it can cause a precipitous drop in the availability of boron to plants as the plants cannot draw nutrients from that desiccated layer. Hence, boron deficiency diseases appear in dry weather.

Boron has many functions within a plant: it affects flowering and fruiting, pollen germination, cell division, and active salt absorption. The metabolism of amino acids and proteins, carbohydrates, calcium, and water are strongly affected by boron. Many of those listed functions may be embodied by its function in moving the highly polar sugars through cell membranes by reducing their polarity and hence the energy needed to pass the sugar. If sugar cannot pass to the fastest growing parts rapidly enough, those parts die. Boron is relatively immobile within a plant suggesting that the molecule is fixed to the points in the membrane where they facilitate sugar transport.

Boron is not relocatable in the plant via the phloem. It must be supplied to the growing parts via the xylem. Foliar sprays affect only those parts sprayed, which may be insufficient for the fastest growing parts, and is very temporary.

Boron is essential for the proper forming and strengthening of cell walls. Lack of boron results in short thick cells producing stunted fruiting bodies and roots. Calcium to boron ratio must be maintained in a narrow range for normal plant growth. For alfalfa, that calcium to boron ratio must be from 80:1 to 600:1. Boron deficiency appears at 800:1 and higher. For alfalfa, similar ratios exist for magnesium, copper, nitrogen and potassium. Boron levels within plants differ with plant species and range from 2.3 p.p.m for barley to 94.7 p.p.m for poppy . Lack of boron causes failure of calcium metabolism which produces hollow heart in beets and peanuts.

Inadequate amounts of boron affect many agricultural crops, legume forage crops most strongly. Of the micronutrients, boron deficiencies are second most common after zinc. Deficiencies of boron when soil is cropped are common and require the application of mineral supplement; one ton of alfalfa hay carries with it one ounce of boron, 100 bushels of peaches 4 ounces. Deficiency results in the death of the terminal growing points. Symptoms first appear as stunted growth, then to cellular changes, which leads to physical changes, and finally death of the plant.

Boron supplements derive from dry lake bed deposits such as those in Death Valley, USA, in the form of sodium tetraborate, from which less soluble calcium borate is made. Foliar sprays are used on fruit crop trees in soils of high alkalinity. Boron is often applied to fields as a contaminant in other soil amendments but is not generally adequate to make up the rate of loss by cropping. The rates of application of borate to produce an adequate alfalfa crop range from 15 pounds per acre for a sandy-silt, acidic soil of low organic matter, to 60 pounds per acre for a soil with high organic matter, high cation exchange capacity and high pH.

Boron concentration in soil water solution higher than one ppm is toxic to most plants. Toxic concentrations within plants are 10 to 50 ppm for small grains and 200 ppm in boron-tolerant crops such as sugar beets, rutabaga, cucumbers, and conifers. Toxic soil conditions are generally limited to arid regions or can be caused by underground borax deposits in contact with water or volcanic gases dissolved in percolating water. Application rates should be limited to a few pounds per acre in a test plot to determine if boron is needed generally. Otherwise, testing for boron levels in plant material is required to determine remedies. Excess boron can be removed by irrigation and assisted by application of elemental sulfur to lower the pH and increase boron's solubility. Application of calcium will increase soil alkalinity, causing boron to fix on the mineral soil component and remove some fraction, thereby reducing boron toxicity.

Boron deficiencies must be detected by analysis of plant material to apply a correction before the obvious symptoms appear, after which it is too late to prevent crop loss.

Strawberries deficient in boron will produce lumpy fruit; apricots will not blossom or, if they do, will not fruit or will drop their fruit depending on the level of boron deficit. Broadcast of boron supplements is effective and long term; a foliar spray is immediate but must be repeated.

Boron is an essential element for the health of animals which derive their boron from plant material.

Copper

Copper is important for photosynthesis. Symptoms for copper deficiency include chlorosis.It is involved in many enzyme processes; necessary for proper photosynthesis; involved in the manufacture of lignin (cell walls) and involved in grain production. It is also hard to find in some soil conditions.

Manganese

Manganese is necessary for photosynthesis, including the building of chloroplasts. Manganese deficiency may result in coloration abnormalities, such as discolored spots on the foliage.

Sodium

Sodium is involved in the regeneration of phosphoenolpyruvate in CAM and C4 plants. Sodium can potentially replace potassium's regulation of stomatal opening and closing.

Essentiality of sodium:

- Essential for C4 plants rather C3

- Substitution of K by Na: Plants can be classified into four groups:

1. Group A—a high proportion of K can be replaced by Na and stimulate the growth, which cannot be achieved by the application of K

2. Group B—specific growth responses to Na are observed but they are much less distinct

3. Group C—Only minor substitution is possible and Na has no effect

4. Group D—No substitution occurs

- Stimulate the growth—increase leaf area and stomata. Improves the water balance

- Na functions in metabolism

1. C4 metabolism

2. Impair the conversion of pyruvate to phosphoenol-pyruvate

3. Reduce the photosystem II activity and ultrastructural changes in mesophyll chloroplast

- Replacing K functions

1. Internal osmoticum

2. Stomatal function

3. Photosynthesis

4. Counteraction in long distance transport

5. Enzyme activation

- Improves the crop quality e.g. improves the taste of carrots by increasing sucrose

Zinc

Zinc is required in a large number of enzymes and plays an essential role in DNA transcription. A typical symptom of zinc deficiency is the stunted growth of leaves, commonly known as "little leaf" and is caused by the oxidative degradation of the growth hormone auxin.

Nickel

In higher plants, nickel is absorbed by plants in the form of Ni^{2+} ion. Nickel is essential for activation of urease, an enzyme involved with nitrogen metabolism that is required to process urea. Without nickel, toxic levels of urea accumulate, leading to the formation of necrotic lesions. In lower plants, nickel activates several enzymes involved in a variety of processes, and can substitute for zinc and iron as a cofactor in some enzymes.

Chlorine

Chlorine, as compounded chloride, is necessary for osmosis and ionic balance; it also plays a role in photosynthesis.

Cobalt

Cobalt has proven to be beneficial to at least some plants although it does not appear to be essential for most species. It has, however, been shown to be essential for nitrogen fixation by the nitrogen-fixing bacteria associated with legumes and other plants.

Aluminium

- Tea has a high tolerance for aluminum (Al) toxicity and the growth is stimulated by Al application. The possible reason is the prevention of Cu, Mn or P toxicity effects.

- There have been reports that Al may serve as a fungicide against certain types of root rot.

Silicon

Silicon is not considered an essential element for plant growth and development. It is always found in abundance in the environment and hence if needed it is available. It is found in the structures of plants and improves the health of plants.

In plants, silicon has been shown in experiments to strengthen cell walls, improve plant strength, health, and productivity. There have been studies showing evidence of silicon improving drought and frost resistance, decreasing lodging potential and boosting the plant's natural pest and disease fighting systems. Silicon has also been shown to improve plant vigor and physiology by improving root mass and density, and increasing above ground plant biomass and crop yields. Silicon is currently under consideration by the Association of American Plant Food Control Officials (AAPFCO) for elevation to the status of a "plant beneficial substance".

Higher plants differ characteristically in their capacity to take up silicon. Depending on their SiO_2 content they can be divided into three major groups:

- Wetland graminae-wetland rice, horsetail (10–15%)

- Dryland graminae-sugar cane, most of the cereal species and few dicotyledons species (1–3%)

- Most of dicotyledons especially legumes (<0.5%)

- The long distance transport of Si in plants is confined to the xylem. Its distribution within the shoot organ is therefore determined by transpiration rate in the organs

- The epidermal cell walls are impregnated with a film layer of silicon and effective barrier against water loss, cuticular transpiration rate in the organs.

Vanadium

Vanadium may be required by some plants, but at very low concentrations. It may also be substituting for molybdenum.

Selenium

Selenium is probably not essential for flowering plants, but it can be beneficial; it can stimulate plant growth, improve tolerance of oxidative stress, and increase resistance to pathogens and herbivory.

Selenium is, however, an essential mineral element for animal (including human) nutrition and selenium deficiencies are known to occur when food or animal feed is grown on selenium-deficient soils. The use of inorganic selenium fertilizers can increase selenium concentrations in edible crops and animal diets thereby improving animal health.

Nutrient Deficiency

The effect of a nutrient deficiency can vary from a subtle depression of growth rate to obvious stunting, deformity, discoloration, distress, and even death. Visual symptoms distinctive enough to be useful in identifying a deficiency are rare. Most deficiencies are multiple and moderate. However, while a deficiency is seldom that of a single nutrient, nitrogen is commonly the nutrient in shortest supply.

Chlorosis of foliage is not always due to mineral nutrient deficiency. Solarization can produce superficially similar effects, though mineral deficiency tends to cause premature defoliation, whereas solarization does not, nor does solarization depress nitrogen concentration.

Nutrient Status of Plants

Nutrient status (mineral nutrient and trace element composition, also called ionome and nutrient profile) of plants are commonly portrayed by tissue elementary analysis. Interpretation of the results of such studies, however, has been controversial. During the last decades the nearly two-century-old "law of minimum" or "Liebig's law" (that states that plant growth is controlled not by the total amount of resources available, but by the scarcest resource) has been replaced by several mathematical approaches that use different models in order to take the interactions between the individual nutrients into account. The latest developments in this field are based on the fact that the nutrient elements (and compounds) do not act independently from each other; Baxter, 2015, because there may be direct chemical interactions between them or they may influence each other's uptake, translocation, and biological action via a number of mechanisms as exemplified for the case of ammonia.

Plant Nutrition in Agricultural Systems

Hydroponics

Hydroponics is a method for growing plants in a water-nutrient solution without the use of nutrient-rich soil. It allows researchers and home gardeners to grow their plants in a controlled environment. The most common solution is the Hoagland solution, developed by D. R. Hoagland in 1933. The solution consists of all the essential nutrients in the correct proportions necessary for most plant growth. An aerator is used to prevent an anoxic event or hypoxia. Hypoxia can affect nutrient uptake of a plant because, without oxygen present, respiration becomes inhibited within the root cells. The nutrient film technique is a hydroponic technique in which the roots are not fully submerged. This allows for adequate aeration of the roots, while a "film" thin layer of nutrient-rich water is pumped through the system to provide nutrients and water to the plant.

Microbial Inoculant

Microbial inoculants also known as soil inoculants are agricultural amendments that use beneficial endophytes (microbes) to promote plant health. Many of the microbes involved form symbiotic relationships with the target crops where both parties benefit (mutualism). While microbial inoculants are applied to improve plant nutrition, they can also be used to promote plant growth by stimulating plant hormone production (Bashan & Holguin, 1997; Sullivan, 2001).

Research into the benefits of inoculants in agriculture extends beyond their capacity as biofertilizers. Microbial inoculants can induce systemic acquired resistance (SAR) of crop species to several common crop diseases (provides resistance against pathogens). So far SAR has been demonstrated for powdery mildew (*Blumeria graminis* f. sp. *hordei*, Heitefuss, 2001), take-all (*Gaeumannomyces graminis* var. *tritici*, Khaosaad *et al.*, 2007), leaf spot (*Pseudomonas syringae*, Ramos Solano *et al.*, 2008) and root rot (*Fusarium culmorum*, Waller *et al.* 2005).

Bacterial

Rhizobacterial Inoculants

The rhizobacteria commonly applied as inoculants include nitrogen-fixers and phosphate-solubilisers which enhance the availability of the macronutrients nitrogen and phosphorus to the host plant. Such bacteria are commonly referred to as plant growth promoting rhizobacteria (PGPR).

Nitrogen-fixing Bacteria

The most commonly applied rhizobacteria are *Rhizobium* and closely related genera. *Rhizobium* are nitrogen-fixing bacteria that form symbiotic associations within nodules on the roots of legumes. This increases host nitrogen nutrition and is important to the cultivation of soybeans, chickpeas and many other leguminous crops. For non-leguminous crops, *Azospirillum* has been demonstrated to be beneficial in some cases for nitrogen fixation and plant nutrition (Bashan & Holguin, 1997).

For cereal crops, diazotrophic rhizobacteria have increased plant growth (Galal *et al.*, 2003), grain yield (Caballero-Mellado *et al.*, 1992), nitrogen and phosphorus uptake (Galal *et al.*, 2003), and nitrogen (Caballero-Mellado *et al.*, 1992), phosphorus (Caballero-Mellado *et al.*, 1992; Belimov *et al.*, 1995) and potassium content (Caballero-Mellado *et al.*, 1992). Rhizobacteria live in root nodes, and are associated with legumes.

Phosphate-solubilising Bacteria

To improve phosphorus nutrition, the use of phosphate-solubilising bacteria (PSB) such as *Agrobacterium radiobacter* has also received attention (Belimov *et al.*, 1995a; 1995b; Singh & Kapoor, 1999). As the name suggests, PSB are free-living bacteria that break down inorganic soil phosphates to simpler forms that enable uptake by plants.

Fungal Inoculants

Several different fungal inoculants have been explored for their benefits to plant nutrition. The most commonly investigated fungi for this purpose are the arbuscular mycorrhizae (AM). Other endophytic fungi, such as *Piriformis indica* can also be beneficial (Waller *et al.*, 2005).

Composite Inoculants

The combination of strains of Plant Growth Promoting Rhizobacteria has been shown to benefit rice (*Oryza*, Nguyen *et al.* (2002)) and barley (*Hordeum*, Belimov *et al.* (1995a)). The main benefit from dual inoculants is increased plant nutrient uptake, from both soil and fertiliser (Bashan *et al.*, 2004; Belimov *et al.* 1995a). Interestingly, multiple strain inoculants have also been demonstrated to increase total nitrogenase activity compared to single strain inoculants, even when only one strain is diazotrophic (Lippi *et al.*, 1992; Khammas & Kaiser, 1992, Belimov *et al.* 1995a).

PGPR and arbuscular mycorrhizae in combination can be useful in increasing wheat growth in nutrient poor soil (Singh & Kapoor, 1999) and improving nitrogen-extraction from fertilised soils (Galal *et al.*, 2003). In salinised soils, Rabie (2005) found that inoculating AM-infected *Vicia faba* plants with *Azospirillum brasilense* amplified the beneficial effects of AM inoculation.

Hoagland Solution

The Hoagland solution is a hydroponic nutrient solution that was developed by Hoagland and Arnon in 1938 and revised by Arnon in 1950 and is one of the most popular solution compositions for growing plants (in the scientific world at least). The Hoagland solution provides every nutrient necessary for plant growth and is appropriate for the growth of a large variety of plant species. The solution described by Hoagland and Arnon in 1950 has been modified several times, mainly to add iron chelates, the original concentrations for each element are shown below.

- N 210 ppm
- K 235 ppm
- Ca 200 ppm
- P 31 ppm
- S 64 ppm
- Mg 48 ppm
- B 0.5 ppm
- Fe 1 to 5 ppm
- Mn 0.5 ppm
- Zn 0.05 ppm
- Cu 0.02 ppm
- Mo 0.01 ppm

The Hoagland solution has a lot of N and K so it is very well suited for the development of large plants like tomato and bell pepper. The solution is very good for the growth of plants with lower nutrient demands as well, such as lettuce and aquatic plants with the further dilution of the preparation to 1/4 or 1/5 of the original.

1. Potassium nitrate(KNO_3)

2. Magnesium sulphate heptahydrate, $MgSO_4 \cdot 7H_2O$, and potassium dihydrogen phosphate (Potassium phosphate monobasic), KH_2PO_4

3. Iron EDTA or Iron chelate, Fe-EDTA

4. Boric Acid, H_3BO_3

5. Copper Sulfate, $CuSO_4$

6. Zinc sulfate heptahydrate, $ZnSO_4 \cdot 7H_2O$

7. Manganese chloride, $MnCl_2 \cdot 4H_2O$

8. Sodium molybdate, $Na_2MoO_4 \cdot 2H_2O$,

9. Calcium nitrate, $Ca(NO_3)_2 \cdot 4H_2O$. Last solution must be added as final.

Component	Stock Solution	mL Stock Solution/1L
Macronutrients		
2M KNO_3	202 g/L	2.5
1M $Ca(NO_3)_2 \cdot 4H_2O$	236 g/0.5L	2.5
Iron (Sprint 138 iron chelate)	15 g/L	1.5
2M $MgSO_4 \cdot 7H_2O$	493 g/L	1
1M NH_4NO_3	80 g/L	1
Micronutrients		
H_3BO_3	2.86 g/L	1
$MnCl_2 \cdot 4H_2O$	1.81 g/L	1
$ZnSO_4 \cdot 7H_2O$	0.22 g/L	1
$CuSO_4 \cdot 5H_2O$	0.051 g/L	1
$H_2MoO_4 \cdot H_2O$ or	0.09 g/L	1
$Na_2MoO_4 \cdot 2H_2O$	0.12 g/L	1
Phosphate		
1M KH_2PO_4 (pH to 6.0)	136 g/L	0.5

Procedure:

1. Make up stock solutions and store in separate bottles with appropriate label.

2. Add each component to 800 mL deionized water then fill to 1 L.

3. After the solution is mixed, it is ready to water plants.

What many persons may not know is that the Hoagland/Arnon nutrient solution formulations have a use intent that one gallon of nutrient solution is used per plant with replacement on a weekly basis. If any of these use parameters are changed, i.e., the volume of solution, number of plants, and/or frequency of replacement, plant performance will be significantly affected, this is a factor that may not be realized by persons using the formulations.

Nutrient Management

Nutrient management is the science and art directed to link soil, crop, weather, and hydrologic factors with cultural, irrigation, and soil and water conservation practices to achieve the goals of optimizing nutrient use efficiency, yields, crop quality, and economic returns, while reducing off-site transport of nutrients that may impact the environment. Nutrient management is the skillful task of matching a specific field soil, climate, and crop management conditions to rate, source, timing, and place (commonly known as the 4R nutrient stewardship) of nutrient application.

Nitrogen fertilizer being applied to growing corn (maize) in a contoured, no-tilled field in Iowa.

Some important factors that need to be considered when managing nutrients include (a) the application of nutrients considering the achievable optimum yields and, in some cases, crop quality; (b) the management, application, and timing of nutrients using a budget based on all sources and sinks active at the site; and (c) the management of soil, water, and crop to minimize the off-site transport of nutrients from nutrient leaching out of the root zone, surface runoff, and volatilization (or other gas exchanges).

There can be potential interactions because of differences in nutrient pathways and dynamics. For instance, practices that reduce the off-site surface transport of a given nutrient may increase the leaching losses of other nutrients. These complex dynamics present nutrient managers the difficult task of integrating soil, crop, weather, hydrology, and management practices to achieve the best balance for maximizing profit while contributing to the conservation of our biosphere.

Nutrient Management Plan

Manure spreader

A crop nutrient management plan is a tool that farmers can use to increase the efficiency of all the nutrient sources a crop uses while reducing production and environmental risk, ultimately increasing profit. It is generally agreed that there are ten fundamental components of a Crop Nutrient Management Plan. Each component is critical to helping analyze each field and improve nutrient efficiency for the crops grown. These components include:

- *Field Map:* The map, including general reference points (such as streams, residences, wellheads etc.), number of acres, and soil types is the base for the rest of the plan.

- *Soil Test:* How much of each nutrient (N-P-K and other critical elements such as pH and organic matter) is in the soil profile? The soil test is a key component needed for developing the nutrient rate recommendation.

- *Crop Sequence:* Did the crop that grew in the field last year (and in many cases two or more years ago) fix nitrogen for use in the following years? Has long-term no-till increased organic matter? Did the end-of-season stalk test show a nutrient deficiency? These factors also need to be factored into your plan.

- *Estimated Yield:* Factors that affect yield are numerous and complex. A field's soils, drainage, insect, weed and disease pressure, rotation and many other factors differentiate one field from another. This is why using historic yields is important in developing yield estimates for next year. Accurate yield estimates can dramatically improve nutrient use efficiency.

- *Sources and Forms:* The sources and forms of available nutrients can vary from farm-to-farm and even field-to-field. For instance, manure fertility analysis, storage practices and other factors will need to be included in a nutrient management plan. Manure nutrient tests/analysis are one way to determine the fertility of it. Nitrogen fixed from a previous year's legume crop and residual affects of manure also effects rate recommendations. Many other nutrient sources should also be factored into this plan.

- *Sensitive Areas:* What's out of the ordinary about a field's plan? Is it irrigated? Next to a stream or lake? Especially sandy in one area? Steep slope or low area? Manure applied in one area for generations due to proximity of dairy barn? Extremely productive—or unproductive—in a portion of the field? Are there buffers that protect streams, drainage ditches, wellheads, and other water collection points? How far away are the neighbors? What's the general wind direction? This is the place to note these and other special conditions that need to be considered.

- *Recommended rates:* Here's the place where science, technology, and art meet. Given everything you've noted, what is the optimum rate of N, P, K, lime and any other nutrients? While science tells us that a crop has changing nutrient requirements during the growing season, a combination of technology and farmer's management skills assure optimum nutrient availability at all stages of growth. No-till corn generally requires starter fertilizer to give the seedling a healthy start.

- *Recommended timing:* When does the soil temperature drop below 50 degrees? Will a N stabilizer be used? What's the tillage practice? Strip-till corn and no-till often require different timing approaches than seed planted into a field that's been tilled once with a field cultivator. Will a starter fertilizer be used to give the seedling a healthy start? How many acres can be covered with available labor (custom or hired) and equipment? Does manure application in a farm depend on a custom applicator's schedule? What agreements have been worked out with neighbors for manure use on their fields? Is a neighbor hosting a special event? All these factors and more will likely figure into the recommended timing.

- *Recommended methods:* Surface or injected? While injection is clearly preferred, there may be situations where injection is not feasible (i.e. pasture, grassland). Slope, rainfall patterns, soil type, crop rotation and many other factors determine which method is best for optimizing nutrient efficiency (availability and loss) in farms. The combination that's right in one field may differ in another field even with the same crop.

- *Annual review and update:* Even the best managers are forced to deviate from their plans. What rate was actually applied? Where? Using which method? Did an unusually mild winter or wet spring reduce soil nitrate? Did a dry summer, disease, or some other unusual factor increase nutrient carryover? These and other factors should be noted. It's easier to make notes throughout the year than to remember back six to 10 months.

When such a plan is designed for animal feeding operations (AFO), it may be termed a "manure management plan." In the United States, some regulatory agencies recommend or require that farms implement these plans in order to prevent water pollution. The U.S. Natural Resources Conservation Service (NRCS) has published guidance documents on preparing a comprehensive nutrient management plan (CNMP) for AFOs.

The International Plant Nutrition Institute has published a 4R plant nutrition manual for improving the management of plant nutrition. The manual outlines the scientific principles behind each of the four R's or "rights" and discusses the adoption of 4R practices on the farm, approaches to nutrient management planning, and measurement of sustainability performance.

Nutrient Budgeting

Nutrient budgets offer insight into the balance between crop inputs and outputs. In short, they compare nutrients you apply to the soil to nutrients taken up by crops. A nutrient budget takes into account all the nutrient inputs on a farm and all those removed from the land. The most obvious source of nutrients in this situation is fertilizer, but this is only part of the picture. Other inputs come with rainfall, in supplements brought on to the farm and in effluent - either farm or dairy factory - spread on the land. In addition, nutrients can be moved around the farm - from an area used for growing silage to the area used to feed it out, from paddock to raceway, and within paddocks in dung and urine patches. Nutrients are removed from the farm in stock sold on, products (meat, milk, wool), crops sold or fed out off farm, and through processes such as nitrate leaching, volatilization and phosphate run-off etc.

Why is it Important?

An accurate nutrient budget is an important tool to provide an early indication of potential problems arising from (i) a nutrient surplus (inputs>outputs), leading to an accumulation of nutrients and increased risk of loss or (ii) a deficit (outputs>inputs), depleting nutrient reserves and increasing the risk of deficiencies and reduced crop yields. They also provide regulatory authorities with a readily-determined, comparative indicator of environmental impact. Overall, nutrient budgets help ensure that farming practices are conducted in an efficient, economic, and environmentally sustainable manner. As such, nutrient budgets are being increasingly required by councils, producer bodies and international markets.

What does it Include?

A nutrient budget isn't as exact as a financial statement. An assortment of variables affects each tract of land. For example, some areas may have had too much manure applied over time or it may have been unevenly distributed. Previous flooding could throw things off, too. It's normal to incorporate limits and assumptions when compiling your budget including the average nutrient removal coefficient values if you don't have them specific to your field.

Soil test: This component is complementary to the budget and lets you know what nutrients are already available to crops and helps you plan input purchases. It is a critical best management practice (BMP) in the 4R strategy.

Yield history: By examining the historical yields of crops take from specific fields, you can calculate nutrient removal over time. Yield history may also help better predict the amount of uptake that will occur with similar crops planted in the future.

Previous applications: Knowing what's been applied to the field in years past will offer insight into what may already be in the ground or what nutrients may no longer be present.

Water: Consider what kind of water has been applied to the field. Does irrigation water contain dissolved nutrients such as nitrogen (N), sulfur (S), or chloride (Cl)? If so, it should be counted as input.

What's around you?: Consider water sources that could run into your field. Is there a manufacturing facility nearby? What makes up these water sources can impact how you plant.

Nutrient Film Technique

Plants placed into nutrient-rich water channels in an NFT system

Nutrient film technique (NFT) is a hydroponic technique wherein a very shallow stream of water containing all the dissolved nutrients required for plant growth is re-circulated past the bare roots of plants in a watertight gully, also known as channels. NFT was developed in the mid 1960s in England by Dr. Alen Cooper. In an ideal system, the depth of the recirculating stream should be very shallow, little more than a film of water, hence the name 'nutrient film'. This ensures that the thick root mat, which develops in the bottom of the channel, has an upper surface, which, although moist, is in the air. Subsequent to this, an abundant supply of oxygen is provided to the roots of the plants. A properly designed NFT system is based on using the right channel slope, the right flow rate, and the right channel length. The main advantage of the NFT system over other forms of hydroponics is that the plant roots are exposed to adequate supplies of water, oxygen and nutrients. In earlier production systems, there was a conflict between the supply of these requirements, since excessive or deficient amounts of one results in an imbalance of one or both of the others. NFT, because of its design, provides a system wherein all three requirements for healthy plant growth can be met at the same time, provided that the simple concept of NFT is always remembered and practiced. The result of these advantages is that higher yields of high-quality produce are obtained over an extended period of cropping. A downside of NFT is that it has very little buffering against interruptions in the flow, e.g., power outages, but, overall, it is one of the more productive techniques.

The same design characteristics apply to all conventional NFT systems. While slopes along channels of 1:100 have been recommended, in practice it is difficult to build a base for channels that is sufficiently true to enable nutrient films to flow without ponding in locally depressed areas. As a consequence, it is recommended that slopes of 1:30 to 1:40 be used. This allows for minor irregularities in the surface, but, even with these slopes, ponding and water logging may occur. The slope may be provided by the floor, or benches or racks may hold the channels and provide the required slope. Both methods are used and depend on local requirements, often determined by the site and crop requirements.

As a general guide, flow rates for each gully should be 1 litre per minute. At planting, rates may be half this, and the upper limit of 2L/min appears about the maximum. Flow rates beyond these extremes are often associated with nutritional problems. Depressed growth rates of many crops have been observed when channels exceed 12 metres in length. On rapidly growing crops, tests have indicated that, while oxygen levels remain adequate, nitrogen may be depleted over the length of the gully. As a consequence, channel length should not exceed 10–15 metres. In situations where this is not possible, the reductions in growth can be eliminated by placing another nutrient feed halfway along the gully and reducing flow rates to 1L/min through each outlet. Care needs to be taken to maintain hygienic conditions and to avoid heavy metal contamination of NFT systems by using mainly plastic or stainless steel pumps and components.

A home-built NFT hydroponic system

A leading protagonist of NFT was Dr.Alan Cooper, a scientist at the Glasshouse Crops Research Station in England who published the book "The ABC of NFT" (Grower Books, London, UK, 1979, 1844pps, Reprinted by Casper Press, Narrabeen, Australia). NFT systems were used by a significant proportion of commercial growers in the UK through the 1980-1990 period but were only used for lettuce in Europe. Dutch growers particularly rejected NFT because of the perceived high risk of disease spread by the recirculating solution. NFT ensures that plants have unlimited access to water at all times, but it is now recognized that fruiting crops can benefit from carefully limited water supplies. Leafy crops like lettuce benefit from unlimited water supplies and are still widely grown using NFT, but now most commercial greenhouse crops of tomatoes, capsicums and cucumbers are grown hydroponically using some kind of inert media, with rockwool being the most important media worldwide. NFT remains a very popular system for home use.

Potato Minitubers

Most potato varieties are maintained in plant tissue culture and micropropagation methods are used to increase the amount of planting material. Since tissue culture plants perform poorly when planted into field soil, they are instead planted into greenhouses or screenhouses to generate tubers, which are referred to as minitubers. In many countries, it is common for NFT or aeroponic systems to be used for production of minitubers from tissue culture plantlets. The minitubers are planted into the field 6 to 14 months after harvest to grow a crop of potatoes. This first crop of field-grown potatoes is typically replanted to generate more potatoes rather than consumed.

References

- Allen V. Barker; D. J. Pilbeam (2007). Handbook of plant nutrition. CRC Press. ISBN 978-0-8247-5904-9. Retrieved 17 August 2010.

- Marschner, Petra, ed. (2012). Marschner's mineral nutrition of higher plants (3rd ed.). Amsterdam: Elsevier/Academic Press. ISBN 9780123849052.

- Norman P. A. Huner; William Hopkins. "3 & 4". Introduction to Plant Physiology 4th Edition. John Wiley & Sons, Inc. ISBN 978-0-470-24766-2.

- Barker, AV; Pilbeam, DJ (2015). Handbook of Plant Nutrition. (2nd ed.). CRC Press. ISBN 9781439881972. Retrieved 5 June 2016.

- White, Philip J. (2016). "Selenium accumulation by plants". Annals of Botany. 117: 217–235. doi:10.1093/aob/mcv180. Retrieved 5 June 2016.

- "Hydroponics 101 - Nutrient Solutions: Historical Background, Formulations and Use, Part 1". Garden Greenhouse. Retrieved 1 October 2014.

- "AAPFCO Board of Directors 2006 Mid-Year Meeting" (PDF). Association of American Plant Food Control Officials. Retrieved 18 July 2011.

- Miranda, Stephen R.; Barker, Bruce (August 4, 2009). "Silicon: Summary of Extraction Methods". Harsco Minerals. Retrieved 18 July 2011.

Plant Hormone: An Integrated Study

Plant hormones help in the growth of the plant, it helps in the regulation of cellular processes in specific cells. Hormones majorly regulate the formation of flowers, stems and leaves. This chapter helps the readers in understanding all the aspects of plant hormones, such as auxin, ethylene, florigen, salicylic acid etc.

Plant Hormone

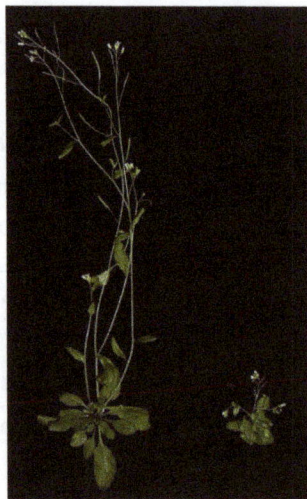

Lack of the plant hormone auxin can cause abnormal growth (right)

Plant hormones (also known as phytohormones) are chemicals that regulate plant growth. In the United Kingdom, these are termed 'plant growth substances'.

Plant hormones are signal molecules produced within the plant, and occur in extremely low concentrations. Hormones regulate cellular processes in targeted cells locally and, moved to other locations, in other functional parts of the plant. Hormones also determine the formation of flowers, stems, leaves, the shedding of leaves, and the development and ripening of fruit. Plants, unlike animals, lack glands that produce and secrete hormones. Instead, each cell is capable of producing hormones. Plant hormones shape the plant, affecting seed growth, time of flowering, the sex of flowers, senescence of leaves, and fruits. They affect which tissues grow upward and which grow downward, leaf formation and stem growth, fruit development and ripening, plant longevity, and even plant death. Hormones are vital to plant growth, and, lacking them, plants would be mostly a mass of undifferentiated cells. So they are also known as growth factors or growth hormones. The term 'Phytohormone' was coined by Thimann in 1948.

Phytohormones are found not only in higher plants, but in algae too, showing similar functions, and in microorganisms, like fungi and bacteria, but, in this case, they play no hormonal or other immediate physiological role in the producing organism and can, thus, be regarded as secondary metabolites.

Characteristics

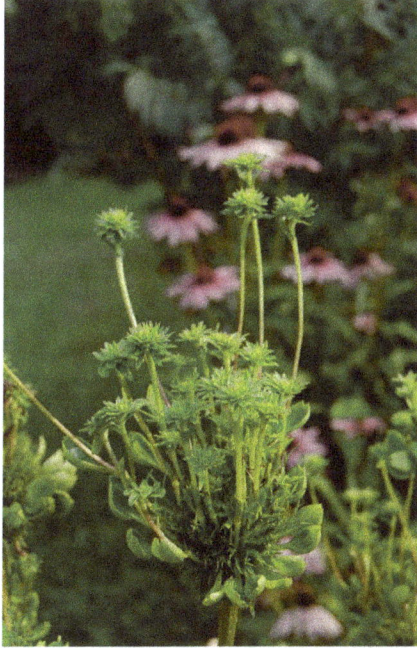

Phyllody on a purple coneflower (*Echinacea purpurea*), a plant development abnormality where leaf-like structures replace flower organs. It can be caused by hormonal imbalance, among other reasons.

The word hormone is derived from Greek, meaning *set in motion*. Plant hormones affect gene expression and transcription levels, cellular division, and growth. They are naturally produced within plants, though very similar chemicals are produced by fungi and bacteria that can also affect plant growth. A large number of related chemical compounds are synthesized by humans. They are used to regulate the growth of cultivated plants, weeds, and in vitro-grown plants and plant cells; these manmade compounds are called Plant Growth Regulators or PGRs for short. Early in the study of plant hormones, "phytohormone" was the commonly used term, but its use is less widely applied now.

Plant hormones are not nutrients, but chemicals that in small amounts promote and influence the growth, development, and differentiation of cells and tissues. The biosynthesis of plant hormones within plant tissues is often diffuse and not always localized. Plants lack glands to produce and store hormones, because, unlike animals — which have two circulatory systems (lymphatic and cardiovascular) powered by a heart that moves fluids around the body — plants use more passive means to move chemicals around the plant. Plants utilize simple chemicals as hormones, which move more easily through the plant's tissues. They are often produced and used on a local basis within the plant body. Plant cells produce hormones that affect even different regions of the cell producing the hormone.

Hormones are transported within the plant by utilizing four types of movements. For localized movement, cytoplasmic streaming within cells and slow diffusion of ions and molecules between cells are utilized. Vascular tissues are used to move hormones from one part of the plant to another; these include sieve tubes or phloem that move sugars from the leaves to the roots and flowers, and xylem that moves water and mineral solutes from the roots to the foliage.

Not all plant cells respond to hormones, but those cells that do are programmed to respond at specific points in their growth cycle. The greatest effects occur at specific stages during the cell's life, with diminished effects occurring before or after this period. Plants need hormones at very specific times during plant growth and at specific locations. They also need to disengage the effects that hormones have when they are no longer needed. The production of hormones occurs very often at sites of active growth within the meristems, before cells have fully differentiated. After production, they are sometimes moved to other parts of the plant, where they cause an immediate effect; or they can be stored in cells to be released later. Plants use different pathways to regulate internal hormone quantities and moderate their effects; they can regulate the amount of chemicals used to biosynthesize hormones. They can store them in cells, inactivate them, or cannibalise already-formed hormones by conjugating them with carbohydrates, amino acids, or peptides. Plants can also break down hormones chemically, effectively destroying them. Plant hormones frequently regulate the concentrations of other plant hormones. Plants also move hormones around the plant diluting their concentrations.

The concentration of hormones required for plant responses are very low (10^{-6} to 10^{-5} mol/L). Because of these low concentrations, it has been very difficult to study plant hormones, and only since the late 1970s have scientists been able to start piecing together their effects and relationships to plant physiology. Much of the early work on plant hormones involved studying plants that were genetically deficient in one or involved the use of tissue-cultured plants grown *in vitro* that were subjected to differing ratios of hormones, and the resultant growth compared. The earliest scientific observation and study dates to the 1880s; the determination and observation of plant hormones and their identification was spread-out over the next 70 years.

Classes of Plant Hormones

In general, it is accepted that there are five major classes of plant hormones, some of which are made up of many different chemicals that can vary in structure from one plant to the next. The chemicals are each grouped together into one of these classes based on their structural similarities and on their effects on plant physiology. Other plant hormones and growth regulators are not easily grouped into these classes; they exist naturally or are synthesized by humans or other organisms, including chemicals that inhibit plant growth or interrupt the physiological processes within plants. Each class has positive as well as inhibitory functions, and most often work in tandem with each other, with varying ratios of one or more interplaying to affect growth regulation.

The five major classes are:

Abscisic Acid Hormone

Abscisic acid (also called ABA) is one of the most important plant growth regulators. It was discovered and researched under two different names before its chemical properties were fully known, it was called *dormin* and *abscicin II*. Once it was determined that the two compounds are the same, it was named abscisic acid. The name "abscisic acid" was given because it was found in high concentrations in newly abscissed or freshly fallen leaves.

This class of PGR is composed of one chemical compound normally produced in the leaves of plants, originating from chloroplasts, especially when plants are under stress. In general, it acts as an inhibitory chemical compound that affects bud growth, and seed and bud dormancy. It mediates changes within the apical meristem, causing bud dormancy and the alteration of the last set of leaves into protective bud covers. Since it was found in freshly abscissed leaves, it was thought to play a role in the processes of natural leaf drop, but further research has disproven this. In plant species from temperate parts of the world, it plays a role in leaf and seed dormancy by inhibiting growth, but, as it is dissipated from seeds or buds, growth begins. In other plants, as ABA levels decrease, growth then commences as gibberellin levels increase. Without ABA, buds and seeds would start to grow during warm periods in winter and be killed when it froze again. Since ABA dissipates slowly from the tissues and its effects take time to be offset by other plant hormones, there is a delay in physiological pathways that provide some protection from premature growth. It accumulates within seeds during fruit maturation, preventing seed germination within the fruit, or seed germination before winter. Abscisic acid's effects are degraded within plant tissues during cold temperatures or by its removal by water washing in out of the tissues, releasing the seeds and buds from dormancy.

In plants under water stress, ABA plays a role in closing the stomata. Soon after plants are water-stressed and the roots are deficient in water, a signal moves up to the leaves, causing the formation of ABA precursors there, which then move to the roots. The roots then release ABA, which is translocated to the foliage through the vascular system and modulates the potassium and sodium uptake within the guard cells, which then lose turgidity, closing the stomata. ABA exists in all parts of the plant and its concentration within any tissue seems to mediate its effects and function as a hormone; its degradation, or more properly catabolism, within the plant affects metabolic reactions and cellular growth and production of other hormones. Plants start life as a seed with high ABA levels. Just before the seed germinates, ABA levels decrease; during germination and early growth of the seedling, ABA levels decrease even more. As plants begin to produce shoots with fully functional leaves, ABA levels begin to increase, slowing down cellular growth in more "mature" areas of the plant. Stress from water or predation affects ABA production and catabolism rates, mediating another cascade of effects that trigger specific responses from targeted cells. Scientists are still piecing together the complex interactions and effects of this and other phytohormones.

Auxins

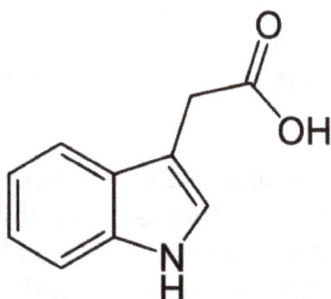

The auxin indole-3-acetic acid

Auxins are compounds that positively influence cell enlargement, bud formation and root initiation. They also promote the production of other hormones and in conjunction with cytokinins, they control the growth of stems, roots, and fruits, and convert stems into flowers. Auxins were the first class of growth regulators discovered. They affect cell elongation by altering cell wall plasticity. They stimulate cambium, a subtype of meristem cells, to divide and in stems cause secondary xylem to differentiate. Auxins act to inhibit the growth of buds lower down the stems (apical dominance), and also to promote lateral and adventitious root development and growth. Leaf abscission is initiated by the growing point of a plant ceasing to produce auxins. Auxins in seeds regulate specific protein synthesis, as they develop within the flower after pollination, causing the flower to develop a fruit to contain the developing seeds. Auxins are toxic to plants in large concentrations; they are most toxic to dicots and less so to monocots. Because of this property, synthetic auxin herbicides including 2,4-D and 2,4,5-T have been developed and used for weed control. Auxins, especially 1-Naphthaleneacetic acid (NAA) and Indole-3-butyric acid (IBA), are also commonly applied to stimulate root growth when taking cuttings of plants. The most common auxin found in plants is indole-3-acetic acid or IAA. The correlation of auxins and cytokinins in the plants is a constant (A/C = const.).

Cytokinins

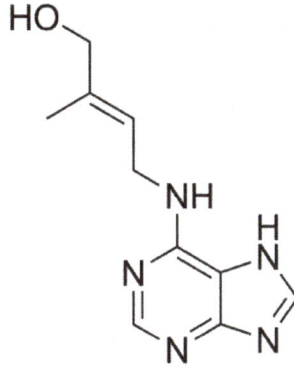

The cytokinin zeatin, the name is derived from *Zea*, in which it was first discovered in immature kernels.

Cytokinins or CKs are a group of chemicals that influence cell division and shoot formation. They were called kinins in the past when the first cytokinins were isolated from yeast cells. They also help delay senescence of tissues, are responsible for mediating auxin transport throughout the plant, and affect internodal length and leaf growth. They have a highly synergistic effect in concert with auxins, and the ratios of these two groups of plant hormones affect most major growth periods during a plant's lifetime. Cytokinins counter the apical dominance induced by auxins; they in conjunction with ethylene promote abscission of leaves, flower parts, and fruits. The correlation of auxins and cytokinins in the plants is a constant (A/C = const.).

Ethylene

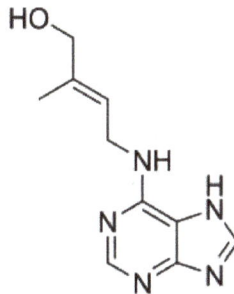

Ethylene

Ethylene is a gas that forms through the breakdown of methionine, which is in all cells. Ethylene has very limited solubility in water and does not accumulate within the cell but diffuses out of the cell and escapes out of the plant. Its effectiveness as a plant hormone is dependent on its rate of production versus its rate of escaping into the atmosphere. Ethylene is produced at a faster rate in rapidly growing and dividing cells, especially in darkness. New growth and newly germinated seedlings produce more ethylene than can escape the plant, which leads to elevated amounts of ethylene, inhibiting leaf expansion. As the new shoot is exposed to light, reactions by phytochrome in the plant's cells produce a signal for ethylene production to decrease, allowing leaf expansion. Ethylene affects cell growth and cell shape; when a growing shoot hits an obstacle while underground, ethylene production greatly increases, preventing cell elongation and causing the stem to swell. The resulting thicker stem can exert more pressure against the object impeding its path to the surface. If the shoot does not reach the surface and the ethylene stimulus becomes prolonged, it affects the stem's natural geotropic response, which is to grow upright, allowing it to grow around an object. Studies seem to indicate that ethylene affects stem diameter and height: When stems of trees are subjected to wind, causing lateral stress, greater ethylene production occurs, resulting in thicker, more sturdy tree trunks and branches. Ethylene affects fruit-ripening: Normally, when the seeds are mature, ethylene production increases and builds-up within the fruit, resulting in a climacteric event just before seed dispersal. The nuclear protein Ethylene Insensitive2 (EIN2) is regulated by ethylene production, and, in turn, regulates other hormones including ABA and stress hormones.

Gibberellins

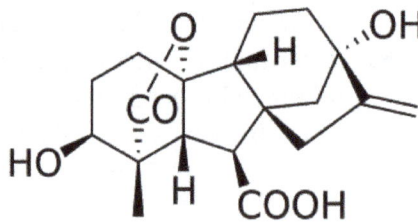

Gibberellin A1

Main function: initiate mobilization of storage materials in seeds during germination, cause elongation of stems, stimulate bolting in biennials stimulate pollen tube growth.

Gibberellins, or GAs, include a large range of chemicals that are produced naturally within plants and by fungi. They were first discovered when Japanese researchers, including Eiichi Kurosawa, noticed a chemical produced by a fungus called *Gibberella fujikuroi* that produced abnormal growth in rice plants. Gibberellins are important in seed germination, affecting enzyme production that mobilizes food production used for growth of new cells. This is done by modulating chromosomal transcription. In grain (rice, wheat, corn, etc.) seeds, a layer of cells called the aleurone layer wraps around the endosperm tissue. Absorption of water by the seed causes production of GA. The GA is transported to the aleurone layer, which responds by producing enzymes that break down stored food reserves within the endosperm, which are utilized by the growing seedling. GAs produce bolting of rosette-forming plants, increasing internodal length. They promote flowering, cellular division, and in seeds growth after germination. Gibberellins also reverse the inhibition of shoot growth and dormancy induced by ABA.

Other Known Hormones

Other identified plant growth regulators include:

- Brassinosteroids - are a class of polyhydroxysteroids, a group of plant growth regulators. Brassinosteroids have been recognized as a sixth class of plant hormones, which stimulate cell elongation and division, gravitropism, resistance to stress, and xylem differentiation. They inhibit root growth and leaf abscission. Brassinolide was the first identified brassinosteroid and was isolated from extracts of rapeseed (*Brassica napus*) pollen in 1979.

- Salicylic acid — activates genes in some plants that produce chemicals that aid in the defense against pathogenic invaders.

- Jasmonates — are produced from fatty acids and seem to promote the production of defense proteins that are used to fend off invading organisms. They are believed to also have a role in seed germination, and affect the storage of protein in seeds, and seem to affect root growth.

- Plant peptide hormones — encompasses all small secreted peptides that are involved in cell-to-cell signaling. These small peptide hormones play crucial roles in plant growth and development, including defense mechanisms, the control of cell division and expansion, and pollen self-incompatibility.

- Polyamines — are strongly basic molecules with low molecular weight that have been found in all organisms studied thus far. They are essential for plant growth and development and affect the process of mitosis and meiosis.

- Nitric oxide (NO) — serves as signal in hormonal and defense responses (e.g. stomatal closure, root development, germination, nitrogen fixation, cell death, stress response). NO can be produced by a yet undefined NO synthase, a special type of nitrite reductase, nitrate reductase, mitochondrial cytochrome c oxidase or non enzymatic processes and regulate plant cell organelle functions (e.g. ATP synthesis in chloroplasts and mitochondria).

- Strigolactones - implicated in the inhibition of shoot branching.

- Karrikins - not plant hormones because they are not made by plants, but are a group of plant growth regulators found in the smoke of burning plant material that have the ability to stimulate the germination of seeds

- Triacontanol - a fatty alcohol that acts as a growth stimulant, especially initiating new basal breaks in the rose family. It is found in alfalfa (lucerne), bee's wax, and some waxy leave cuticles.

Potential Medical Applications

Plant stress hormones activate cellular responses, including cell death, to diverse stress situations in plants. Researchers have found that some plant stress hormones share the ability to adversely affect human cancer cells. For example, sodium salicylate has been found to suppress proliferation of lymphoblastic leukemia, prostate, breast, and melanoma human cancer cells. Jasmonic acid, a plant stress hormone that belongs to the jasmonate family, induced death in lymphoblastic leukemia cells. Methyl jasmonate has been found to induce cell death in a number of cancer cell lines.

Hormones and Plant Propagation

Synthetic plant hormones or PGRs are commonly used in a number of different techniques involving plant propagation from cuttings, grafting, micropropagation, and tissue culture.

The propagation of plants by cuttings of fully developed leaves, stems, or roots is performed by gardeners utilizing auxin as a rooting compound applied to the cut surface; the auxins are taken into the plant and promote root initiation. In grafting, auxin promotes callus tissue formation, which joins the surfaces of the graft together. In micropropagation, different PGRs are used to promote multiplication and then rooting of new plantlets. In the tissue-culturing of plant cells, PGRs are used to produce callus growth, multiplication, and rooting.

Seed Dormancy

Plant hormones affect seed germination and dormancy by acting on different parts of the seed.

Embryo dormancy is characterized by a high ABA:GA ratio, whereas the seed has a high ABA sensitivity and low GA sensitivity. In order to release the seed from this type of dormancy and initiate seed germination, an alteration in hormone biosynthesis and degradation toward a low ABA/GA ratio, along with a decrease in ABA sensitivity and an increase in GA sensitivity, must occur.

ABA controls embryo dormancy, and GA embryo germination. Seed coat dormancy involves the mechanical restriction of the seed coat. This, along with a low embryo growth potential, effectively produces seed dormancy. GA releases this dormancy by increasing the embryo growth potential, and/or weakening the seed coat so the radical of the seedling can break through the seed coat. Different types of seed coats can be made up of living or dead cells, and both types can be influenced by hormones; those composed of living cells are acted upon after seed formation, whereas the seed coats composed of dead cells can be influenced by hormones during the formation of the seed coat. ABA affects testa or seed coat growth characteristics, including thickness, and effects the GA-mediated embryo growth potential. These conditions and effects occur during the formation of the seed, often in response to environmental conditions. Hormones also mediate endosperm dormancy: Endosperm in most seeds is composed of living tissue that can actively respond to hormones generated by the embryo. The endosperm often acts as a barrier to seed germination, playing a part in seed coat dormancy or in the germination process. Living cells respond to and also affect the ABA:GA ratio, and mediate cellular sensitivity; GA thus increases the embryo growth potential and can promote endosperm weakening. GA also affects both ABA-independent and ABA-inhibiting processes within the endosperm.

Abscisic Acid

Abscisic acid (ABA), also known as abscisin II and dormin, is best known as a plant hormone. ABA functions in many plant developmental processes, including bud dormancy. It is degraded by the enzyme (+)-abscisic acid 8'-hydroxylase into phaseic acid.

In Plants

Function

ABA was originally believed to be involved in abscission. This is now known to be the case only in a small number of plants. ABA-mediated signaling also plays an important part in plant responses to environmental stress and plant pathogens. The plant genes for ABA biosynthesis and sequence of the pathway have been elucidated. ABA is also produced by some plant pathogenic fungi via a biosynthetic route different from ABA biosynthesis in plants.

Abscisic acid owes its names to its role in the abscission of plant leaves. In preparation for winter, ABA is produced in terminal buds. This slows plant growth and directs leaf primordia to develop scales to protect the dormant buds during the cold season. ABA also inhibits the division of cells in the vascular cambium, adjusting to cold conditions in the winter by suspending primary and secondary growth.

Abscisic acid is also produced in the roots in response to decreased soil water potential and other situations in which the plant may be under stress. ABA then translocates to the leaves, where it rapidly alters the osmotic potential of stomatal guard cells, causing them to shrink and stomata to close. The ABA-induced stomatal closure reduces transpiration, thus preventing further water loss from the leaves in times of low water availability. A close linear correlation was found between the ABA content of the leaves and their conductance (stomatal resistance) on a leaf area basis.

Seed germination is inhibited by ABA in antagonism with gibberellin. ABA also prevents loss of seed dormancy.

Several ABA-mutant *Arabidopsis thaliana* plants have been identified and are available from the Nottingham Arabidopsis Stock Centre - both those deficient in ABA production and those with altered sensitivity to its action. Plants that are hypersensitive or insensitive to ABA show phenotypes in seed dormancy, germination, stomatal regulation, and some mutants show stunted growth and brown/yellow leaves. These mutants reflect the importance of ABA in seed germination and early embryo development.

Pyrabactin (a pyridyl containing ABA activator) is a naphthalene sulfonamide hypocotyl cell expansion inhibitor, which is an agonist of the seed ABA signaling pathway. It is the first agonist of the ABA pathway that is not structurally related to ABA.

Homeostasis

Biosynthesis

Abscisic acid (ABA) is an isoprenoid plant hormone, which is synthesized in the plastidal 2-C-methyl-D-erythritol-4-phosphate (MEP) pathway; unlike the structurally related sesquiterpenes, which are formed from the mevalonic acid-derived precursor farnesyl diphosphate (FDP), the C_{15} backbone of ABA is formed after cleavage of C_{40} carotenoids in MEP. Zeaxanthin is the first committed ABA precursor; a series of enzyme-catalyzed epoxidations and isomerizations via violaxanthin, and final cleavage of the C_{40} carotenoid by a dioxygenation reaction yields the proximal ABA precursor, xanthoxin, which is then further oxidized to ABA. via abscisic aldehyde.

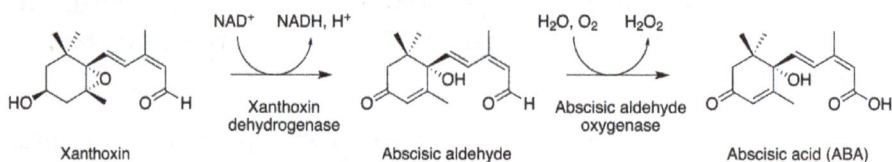

Abamine has been designed, synthesized, developed and then patented as the first specific ABA biosynthesis inhibitor, which makes it possible to regulate endogenous levels of ABA.

Location and Timing of ABA Biosynthesis

- Released during desiccation of the vegetative tissues and when roots encounter soil compaction.

- Synthesized in green fruits at the beginning of the winter period

- Synthesized in maturing seeds, establishing dormancy

- Mobile within the leaf and can be rapidly translocated from the roots to the leaves by the transpiration stream in the xylem

- Produced in response to environmental stress, such as heat stress, water stress, salt stress

- Synthesized in all plant parts, e.g., roots, flowers, leaves and stems

Inactivation

ABA can be catabolized to phaseic acid via CYP707A (a group of P450 enzymes) or inactivated by glucose conjugation (ABA-glucose ester) via the enzyme AOG. Catabolism via the CYP707As is very important for ABA homeostasis, and mutants in those genes generally accumulate higher levels of ABA than lines overexpressing ABA biosynthetic genes. In soil bacteria, an alternative catabolic pathway leading to dehydrovomifoliol via the enzyme vomifoliol dehydrogenase has been reported.

Effects

- Antitranspirant - Induces stomatal closure, decreasing transpiration to prevent water loss.

- Inhibits fruit ripening

- Responsible for seed dormancy by inhibiting cell growth – inhibits seed germination

- Inhibits the synthesis of Kinetin nucleotide

- Downregulates enzymes needed for photosynthesis.

- Acts on endodermis to prevent growth of roots when exposed to salty conditions

- Delays Cell Division

In Fungi

Like plants, some fungal species (for example Cercospora rosicola, Botrytis cinerea and Magnaporthe oryzae) have an endogenous biosynthesis pathway for ABA. In fungi, it seems to be the MVA biosynthetic pathway that is predominant (rather than the MEP pathway that is responsible for ABA biosynthesis in plants). One role of ABA produced by these pathogens seems to be to suppress the plant immune responses.

In Animals

ABA has also been found to be present in metazoans, from sponges up to mammals including humans. Currently, its biosynthesis and biological role in animals is poorly known. ABA has recently been shown to elicit potent anti-inflammatory and anti-diabetic effects in mouse models of diabetes/obesity, inflammatory bowel disease, atherosclerosis and influenza infection. Many biological effects in animals have been studied using ABA as a nutraceutical or pharmacognostic drug, but ABA is also generated endogenously by some cells (like macrophages) when stimulated. There are also conflicting conclusions from different studies, where some claim that ABA is essential for pro-inflammatory responses whereas other show anti-inflammatory effects. Like with many natural substances with medical properties, ABA has become popular also in naturopathy. Whlile ABA clearly has beneficial biological activities and many naturopathic remedies will contain high levels of ABA (such as wheatgrass juice, fruits and vegetables), some of the health claims made may be exaggerated or overly optimistic. Its anti-cancer properties are, for example, poorly supported at this moment but not completely dismissed. (and A1 US application US20060292215 A1, Gonzalo Romero M, "Abscisic acid against cancer", published 2006-12-28). In mammalian cells ABA targets a protein known as lanthionine synthetase C-like 2 (LANCL2), triggering an alternative mechanism of activation of peroxisome prolifer-ator-activated receptor gamma (PPAR gamma). Interestingly, LANCL2 is conserved in plants and was originally suggested to be an ABA receptor also in plants, which was later challenged.

An aquatic herbicide, fluridone, has been found to act as an anti-inflammatory drug in humans. Fluridone inhibits photosynthesis by disruption of ABA, killing plants system-ically. This same inhibition of ABA in humans leads to an anti-inflammatory response.

Oral ABA at 0.5–1 µg/kg significantly lowered hyperglycemia and insulinemia in rats and in humans. So, low-dose ABA intake may be proposed as an aid to improving glu-cose tolerance in patients with diabetes who are deficient in or resistant to insulin.

Measurement of ABA Concentration

Several methods can help to quantify the concentration of abscisic acid in a variety of plant tissue. The quantitative methods used are based on HPLC and GC, and ELISA. Recently, 2 independent FRET probes have been developed that can measure intracel-lular ABA concentrations in real time in vivo.

Salicylic Acid

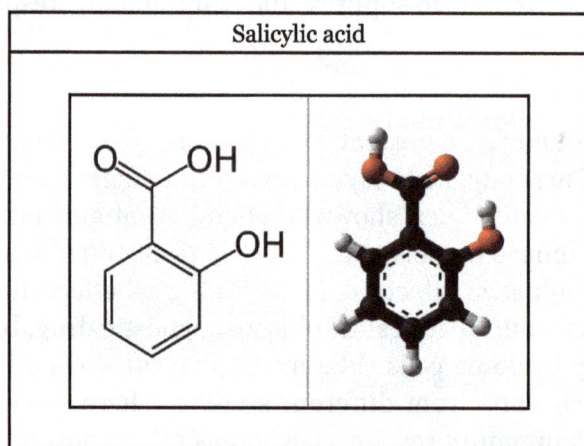

Salicylic acid

Names	
Preferred IUPAC name	
2-Hydroxybenzoic acid	
Identifiers	
CAS Number	69-72-7 ✓
ChEBI	CHEBI:16914 ✓
ChEMBL	ChEMBL424 ✓
ChemSpider	331 ✓
DrugBank	DB00936 ✓
ECHA InfoCard	100.000.648
EC Number	200-712-3
IUPHAR/BPS	4306
Jmol 3D model	Interactive image
KEGG	D00097 ✓
PubChem	338
RTECS number	VO0525000
UNII	O414PZ4LPZ ✓

InChI

- InChI=1S/C7H6O3/c8-6-4-2-1-3-5(6)7(9)10/h1-4,8H,(H,9,10) ✓

Key: YGSDEFSMJLZEOE-UHFFFAOYSA-N ✓

- InChI=1/C7H6O3/c8-6-4-2-1-3-5(6)7(9)10/h1-4,8H,(H,9,10)

Key: YGSDEFSMJLZEOE-UHFFFAOYAQ

SMILES

- c1ccc(c(c1)C(=O)O)O

Properties	
Chemical formula	$C_7H_6O_3$
Molar mass	138.12 g·mol^{-1}
Appearance	colorless to white crystals
Odor	odorless
Density	1.443 g/cm³ (20 °C)

Melting point	158.6 °C (317.5 °F; 431.8 K)
Boiling point	200 °C (392 °F; 473 K) decomposes 211 °C (412 °F; 484 K) at 20 mmHg
Sublimation conditions	sublimes at 76 °C
Solubility in water	1.24 g/L (0 °C) 2.48 g/L (25 °C) 4.14 g/L (40 °C) 17.41 g/L (75 °C) 77.79 g/L (100 °C)
Solubility	soluble in ether, CCl_4, benzene, propanol, acetone, ethanol, oil of turpentine, toluene
Solubility in benzene	0.46 g/100 g (11.7 °C) 0.775 g/100 g (25 °C) 0.991 g/100 g (30.5 °C) 2.38 g/100 g (49.4 °C) 4.4 g/100 g (64.2 °C)
Solubility in chloroform	2.22 g/100 mL (25 °C) 2.31 g/100 mL (30.5 °C)
Solubility in methanol	40.67 g/100 g (−3 °C) 62.48 g/100 g (21 °C)
Solubility in olive oil	2.43 g/100 g (23 °C)
Solubility in acetone	39.6 g/100 g (23 °C)
log P	2.26
Vapor pressure	10.93 mPa
Acidity (pK_a)	1 = 2.97 (25 °C) 2 = 13.82 (20 °C)
UV-vis (λ_{max})	210 nm, 234 nm, 303 nm (4 mg % in ethanol)
Refractive index (n_D)	1.565 (20 °C)
Thermochemistry	
Std enthalpy of formation ($\Delta_f H^\circ_{298}$)	-589.9 kJ/mol
Std enthalpy of combustion ($\Delta_c H^\circ_{298}$)	3.025 MJ/mol
Pharmacology	
ATC code	A01AD05 (WHO) B01AC06 (WHO) D01AE12 (WHO) N02BA01 (WHO) S01BC08 (WHO)
Hazards	
Safety data sheet	MSDS

GHS pictograms	
GHS signal word	Danger
GHS hazard statements	H302, H318
GHS precautionary statements	P280, P305+351+338
EU classification (DSD)	Xn
R-phrases	R22, R38, R41, R61
S-phrases	S22, S26, S36, S37, S39
Eye hazard	Severe irritation
Skin hazard	Mild irritation
NFPA 704	1 2 0
Flash point	157 °C (315 °F; 430 K) closed cup
Autoignition temperature	540 °C (1,004 °F; 813 K)
Lethal dose or concentration (*LD*, *LC*):	
LD_{50} (median dose)	480 mg/kg (mice, oral)
Related compounds	
Related compounds	Methyl salicylate, Benzoic acid, Phenol, Aspirin, 4-Hydroxybenzoic acid, Magnesium salicylate, Choline salicylate, Bismuth subsalicylate, Sulfosalicylic acid
Except where otherwise noted, data are given for materials in their standard state (at 25 °C [77 °F], 100 kPa).	

Salicylic acid is a monohydroxybenzoic acid, a type of phenolic acid and a beta hydroxy acid. It has the formula $C_7H_6O_3$. This colorless crystalline organic acid is widely used in organic synthesis and functions as a plant hormone. It is derived from the metabolism of salicin. In addition to serving as an important active metabolite of aspirin (*acetylsalicylic acid*), which acts in part as a prodrug to salicylic acid, it is probably best known for its use as a key ingredient in topical anti-acne products. The salts and esters of salicylic acid are known as salicylates. The medicinal part of the plant is the inner bark.

It is on the WHO Model List of Essential Medicines, the most important medications needed in a basic health system.

Uses

Medicine

Salicylic acid is known for its ability to ease aches and pains and reduce fevers. These medicinal properties, particularly fever relief, were discovered in ancient times. It is used as an anti-inflammatory drug.

In modern medicine, salicylic acid and its derivatives are constituents of some rubefacient ("skin-reddening") products. For example, methyl salicylate is used as a liniment to soothe joint and muscle pain and choline salicylate is used topically to relieve the pain of mouth ulcers.

Cotton pads soaked in salicylic acid can be used to chemically exfoliate skin

As with other hydroxy acids, salicylic acid is a key ingredient in many skin-care products for the treatment of seborrhoeic dermatitis, acne, psoriasis, calluses, corns, keratosis pilaris, acanthosis nigricans, ichthyosis and warts.

Salicylic acid works as a keratolytic, comedolytic and bacteriostatic agent, causing the cells of the epidermis to shed more readily, opening clogged pores and neutralizing bacteria within, preventing pores from clogging up again and allowing room for new cell growth.

Because of its effect on skin cells, salicylic acid is used in some shampoos to treat dandruff. Concentrated solutions of salicylic acid may cause hyperpigmentation on people with darker skin types (Fitzpatrick phototypes IV, V, VI), without a broad spectrum sunblock.

Bismuth subsalicylate, a salt of bismuth and salicylic acid, is the active ingredient in stomach relief aids such as Pepto-Bismol, is the main ingredient of Kaopectate and "displays anti-inflammatory action (due to salicylic acid) and also acts as an antacid and mild antibiotic".

Salicylic acid is used in the production of other pharmaceuticals, including 4-aminosalicylic acid sandulpiride landetimide (via Salethamide).

In 2016 salicylic acid — and (more potently) diflunisal, a cousin of aspirin — was shown to suppress proteins responsible for cellular damage caused by inflammation.

Chemistry

Salicylic acid has the formula $C_6H_4(OH)COOH$, where the OH group is *ortho* to the carboxyl group. It is also known as 2-hydroxybenzoic acid. It is poorly soluble in water (2 g/L at 20 °C). Aspirin (acetylsalicylic acid or ASA) can be prepared by the esterification of the phenolic hydroxyl group of salicylic acid with the acetyl group from acetic anhydride or acetyl chloride. Salicylic acid can also be prepared using the Kolbe-Schmitt reaction.

Other Uses

Salicylic acid is used as a food preservative, a bactericidal and an antiseptic.

Sodium salicylate is a useful phosphor in the vacuum ultraviolet, with nearly flat quantum efficiency for wavelengths between 10 and 100 nm. It fluoresces in the blue at 420 nm. It is easily prepared on a clean surface by spraying a saturated solution of the salt in methanol followed by evaporation.

Safety

As a topical agent and as a beta-hydroxy acid (and unlike alpha-hydroxy acids), salicylic acid is capable of penetrating and breaking down fats and lipids, causing moderate chemical burns of the skin at very high concentrations. It may damage the lining of pores if the solvent is alcohol, acetone or an oil. Over-the-counter limits are set at 2% for topical preparations expected to be left on the face and 3% for those expected to be washed off, such as acne cleansers or shampoo. 17% and 27% salicylic acid, which is sold for wart removal, should not be applied to the face and should not be used for acne treatment. Even for wart removal, such a solution should be applied once or twice a day – more frequent use may lead to an increase in side-effects without an increase in efficacy.

When ingested, salicylic acid has a possible ototoxic effect by inhibiting prestin. It can induce transient hearing loss in zinc-deficient rats. An injection of salicylic acid induced hearing loss, while an injection of zinc reversed the hearing loss. An injection of magnesium in the zinc-deficient rats did not reverse the induced hearing loss.

No studies examine topical salicylic acid in pregnancy. Oral salicylic acid is not associated with an increase in malformations if used during the first trimester, but in late pregnancy has been associated with bleeding, especially intracranial bleeding. The risks of aspirin late in pregnancy are probably not relevant for a topical exposure to salicylic acid, even late in the pregnancy, because of its low systemic levels. Topical salicylic acid is common in many over-the-counter dermatological agents and the lack of adverse reports suggests a low teratogenic potential.

Salicylic acid overdose can lead to salicylate intoxication, which often presents clinically in a state of metabolic acidosis with compensatory respiratory alkalosis. In patients presenting with an acute overdose, a 16% morbidity rate and a 1% mortality rate are observed.

Some people are hypersensitive to salicylic acid and related compounds.

The United States Food and Drug Administration (FDA) recommends the use of sun protection when using skincare products containing salicylic acid (or any other BHA) on sun-exposed skin areas.

Data support an association between exposure to salicylic acid and Reye's Syndrome. The National Reye's Syndrome Foundation cautions against the use of these and other substances similar to aspirin on children and adolescents. Epidemiological research associated the development of Reye's Syndrome and the use of aspirin for treating the symptoms of influenza-like illnesses, chicken pox, colds, etc. The U.S. Surgeon General, the FDA, the Centers for Disease Control and Prevention and the American Academy of Pediatrics recommend that aspirin and combination products containing aspirin not be given to children under 19 years of age during episodes of fever-causing illnesses.

Plant Hormone

Salicylic acid (SA) is a phenolic phytohormone and is found in plants with roles in plant growth and development, photosynthesis, transpiration, ion uptake and transport. SA also induces specific changes in leaf anatomy and chloroplast structure. SA is involved in endogenous signaling, mediating in plant defense against pathogens. It plays a role in the resistance to pathogens by inducing the production of pathogenesis-related proteins. It is involved in the systemic acquired resistance (SAR) in which a pathogenic attack on one part of the plant induces resistance in other parts. The signal can also move to nearby plants by salicylic acid being converted to the volatile ester, methyl salicylate.

Production

Salicylic acid is biosynthesized from the amino acid phenylalanine. In *Arabidopsis thaliana* it can be synthesized via a phenylalanine-independent pathway.

Sodium salicylate is commercially prepared by treating sodium phenolate (the sodium salt of phenol) with carbon dioxide at high pressure (100 atm) and high temperature (390 K) – a method known as the Kolbe-Schmitt reaction. Acidification of the product with sulfuric acid gives salicylic acid:

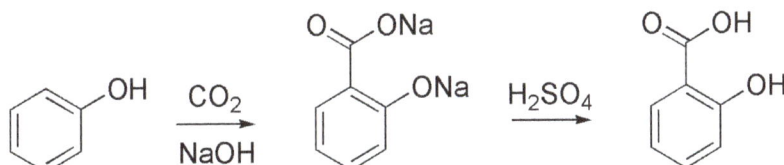

It can also be prepared by the hydrolysis of aspirin (acetylsalicylic acid) or methyl salicylate (oil of wintergreen) with a strong acid or base.

History

Salix alba

White willow (*Salix alba*) is a natural source of salicylic acid

Hippocrates, Galen, Pliny the Elder and others knew that willow bark could ease pain and reduce fevers. It was used in Europe and China to treat these conditions. This remedy is mentioned in texts from ancient Egypt, Sumer and Assyria. The Cherokee and other Native Americans used an infusion of the bark for fever and other medicinal purposes.

In 2014, archaeologists identified traces of salicylic acid on 7th century pottery fragments found in east central Colorado. The Reverend Edward Stone, a vicar from Chipping Norton, Oxfordshire, England, noted in 1763 that the bark of the willow was effective in reducing a fever.

The active extract of the bark, called *salicin*, after the Latin name for the white willow (*Salix alba*), was isolated and named by the German chemist Johann Andreas Buchner in 1828. A larger amount of the substance was isolated in 1829 by Henri Leroux, a French pharmacist. Raffaele Piria, an Italian chemist, was able to convert the substance into a sugar and a second component, which on oxidation becomes salicylic acid.

Salicylic acid was also isolated from the herb meadowsweet (*Filipendula ulmaria*, formerly classified as *Spiraea ulmaria*) by German researchers in 1839. While their extract was somewhat effective, it also caused digestive problems such as gastric irritation, bleeding, diarrhea and even death when consumed in high doses.

Dietary Sources

Unripe fruits and vegetables are natural sources of salicylic acid, particularly blackberries, blueberries, cantaloupes, dates, grapes, kiwi fruits, guavas, apricots, green pepper, olives, tomatoes, radish, chicory and mushrooms. Some herbs and spices contain high amounts, while meat, poultry, fish, eggs and dairy products all have little to no salicylates. Of the legumes, seeds, nuts and cereals, only almonds, water chestnuts and peanuts have significant amounts.

Mechanism of Action

Salicylic acid works through several different pathways. It produces its anti-inflammatory effects via suppressing the activity of cyclooxygenase (COX), an enzyme that is responsible for the production of pro-inflammatory mediators such as the prostaglandins. It does this not by direct inhibition of COX like most other non-steroidal anti-inflammatory drugs (NSAIDs) but instead by suppression of the expression of the enzyme through a yet-unelucidated mechanism.

Salicylic acid activates adenosine monophosphate-activated protein kinase (AMPK) and this action may play a role in the anticancer effects of the compound and its prodrugs aspirin and salsalate. The antidiabetic effects of salicylic acid are likely mediated by AMPK activation primarily through allosteric conformational change that increases levels of phosphorylation.

Salicylic acid also uncouples oxidative phosphorylation, which leads to increased ADP:ATP and AMP:ATP ratios in the cell. As a consequence, salicylic acid may alter AMPK activity and work as an anti-diabetic by altering the energy status of the cell. AMPK knock-out mice display an anti-diabetic effect, demonstrating at least one additional, yet-unidentified action.

Salicylic acid regulates c-Myc level at both transcriptional and post-transcription levels. Inhibition of c-Myc may be an important pathway by which aspirin exerts an anti-cancer effect, decreasing the occurrence of cancer in epithelial tissues.

Auxin

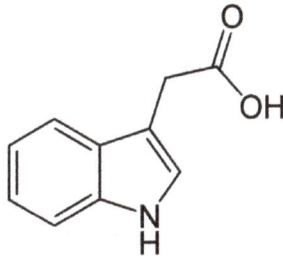

indole-3-acetic acid (IAA) is the most abundant and the basic auxin natively occurring and functioning in plants. It generates the majority of auxin effects in intact plants, and is the most potent native auxin.

There are four more endogenously synthesized auxins in plants.All auxins are compounds with aromatic ring and a carboxylic acid group:

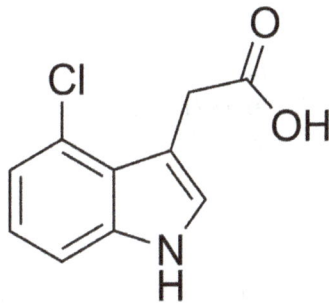

4-Chloroindole-3-acetic acid (4-CI-IAA)

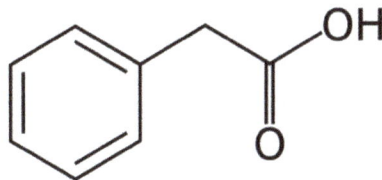

2-phenylacetic acid (PAA)

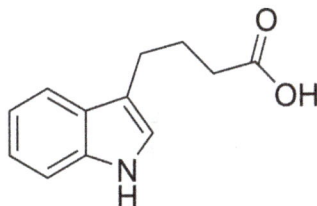

Indole-3-butyric acid (IBA)

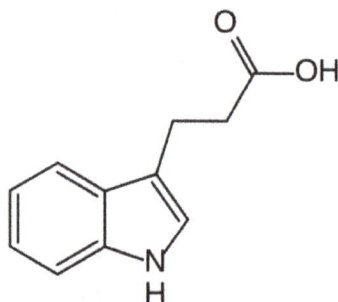

Indole-3-propionic acid (IPA)

Auxins are a class of plant hormones (or plant growth substances) with some morphogen-like characteristics. Auxins have a cardinal role in coordination of many growth and behavioral processes in the plant's life cycle and are essential for plant body development. Auxins and their role in plant growth were first described by the Dutch scientist Frits Warmolt Went. Kenneth V. Thimann was the first to isolate one of these phytohormones and determine its chemical structure as indole-3-acetic acid (IAA). Went and Thimann co-authored a book on plant hormones, *Phytohormones*, in 1937.

Overview

Auxins were the first of the major plant hormones to be discovered. They derive their name from the word (*auxein* – "to grow/increase"). Auxin (namely IAA) is present in all parts of a plant, although in very different concentrations. The concentration in each position is crucial developmental information, so it is subject to tight regulation through both metabolism and transport. The result is the auxin creates "patterns" of auxin concentration maxima and minima in the plant body, which in turn guide further development of respective cells, and ultimately of the plant as a whole.

The (dynamic and environment responsive) pattern of auxin distribution within the plant is a key factor for plant growth, its reaction to its environment, and specifically for development of plant organs (such as leaves or flowers). It is achieved through very complex and well coordinated active transport of auxin molecules from cell to cell throughout the plant body — by the so-called polar auxin transport. Thus, a plant can (as a whole) react to external conditions and adjust to them, without requiring a nervous system. Auxins typically act in concert with, or in opposition to, other plant hormones. For example, the ratio of auxin to cytokinin in certain plant tissues determines initiation of root versus shoot buds.

On the molecular level, all auxins are compounds with an aromatic ring and a carboxylic acid group. The most important member of the auxin family is indole-3-acetic acid (IAA), which generates the majority of auxin effects in intact plants, and is the most potent native auxin. And as native auxin, its stability is controlled in many ways in plants, from synthesis, through possible conjugation to degradation of its molecules, always according to the requirements of the situation. However, molecules of IAA are chemically labile in aqueous solution, so it is not used commercially as a plant growth regulator.

- Five naturally occurring (endogenous) auxins in plants include indole-3-acetic acid, 4-chloroindole-3-acetic acid, phenylacetic acid, indole-3-butyric acid, and indole-3-propionic acid. However, most of the knowledge described so far in auxin biology and as described in the article below, apply basically to IAA; the other three endogenous auxins seems to have rather marginal importance for intact plants in natural environments. Alongside endogenous auxins, scientists and manufacturers have developed many synthetic compounds with auxinic activity.

- Synthetic auxin analogs include 1-naphthaleneacetic acid, 2,4-dichlorophenoxyacetic acid (2,4-D), and many others.

Some synthetic auxins, such as 2,4-D and 2,4,5-trichlorophenoxyacetic acid (2,4,5-T), are used also as herbicides. Broad-leaf plants (dicots), such as dandelions, are much more susceptible to auxins than narrow-leaf plants (monocots) such as grasses and cereal crops, so these synthetic auxins are valuable as synthetic herbicides.

Auxins are also often used to promote initiation of adventitious roots, and are the active ingredient of the commercial preparations used in horticulture to root stem cuttings. They can also be used to promote uniform flowering and fruit set, and to prevent premature fruit drop.

Discovery of Auxin

Charles Darwin

In 1881, Charles Darwin and his son Francis performed experiments on coleoptiles, the sheaths enclosing young leaves in germinating grass seedlings. The experiment exposed the coleoptile to light from a unidirectional source and observed that they bend towards the light. By covering various parts of the coleoptiles with a light impermeable opaque cap, the Darwins discovered that light is detected by the coleoptile tip, but that bending occurs in the hypocotyl. However the seedlings showed no signs of development towards light if the tip was covered with an opaque cap, or if the tip was removed. The Darwins concluded that the tip of the coleoptile was responsible for sensing light, and proposed that a messenger is transmitted in a downward direction from the tip of the coleoptile, causing it to bend.

Peter Boysen-Jensen

In 1913, Danish scientist Peter Boysen-Jensen demonstrated that the signal was not transfixed but mobile. He separated the tip from the remainder of the coleoptile by a cube of gelatine which prevented cellular contact, but allowed chemicals to pass through. The seedlings responded normally bending towards the light. However, when the tip was separated by an impermeable substance, there was no curvature of the stem.

Frits Went

In 1926, the Dutch botanist Frits Warmolt Went showed that a chemical messenger diffuses from coleoptile tips. Went's experiment identified how a growth promoting chemical causes a coleoptile to grow towards light. Went cut the tips of the coleoptiles and placed them in the dark, putting a few tips on agar blocks that he predicted would absorb the growth-promoting chemical. On control coleoptiles, he placed a block that lacked the chemical. On others, he placed blocks containing the chemical, either centred on top of the coleoptile to distribute the chemical evenly or offset to increase the concentration on one side. When the growth promoting chemical was distributed evenly the coleoptile grew straight. If the chemical was distributed unevenly, the coleoptile curved away from the side with the cube, as if growing towards light, even though it was grown in the dark. Went later proposed that the messenger substance is a growth-promoting hormone, which he named auxin, that becomes asymmetrically distributed in the bending region. Went concluded that auxin is at a higher concentration on the shaded side, promoting cell elongation, which results in a coleoptiles bending towards the light.

Hormonal Activity

Auxins help development at all levels in plants, from the cellular level, through organs, and ultimately to the whole plant.

Molecular Mechanisms

Auxin molecules present in cells may trigger responses directly through stimulation or inhibition of the expression of sets of certain genes or by means independent of gene expression. Auxin transcriptionally activates four different families of *early genes* (aka *primary response genes*), so-called because the components required for the activation are preexisting, leading to a rapid response. The families are glutathione S-transferases, auxin homeostasis proteins like GH3, SAUR genes of currently unknown function, and the Aux/IAA repressors.

Aux/IAA, ARF, TIR1, SCF Auxin Regulatory Pathways

The Aux/IAA repressors provide an example of one of the pathways leading to auxin induced changes of gene expression. This pathway involves the protein families TIR1 (transport inhibitor response1), ARF (auxin response factor), Aux/IAA transcriptional repressors, and the ubiquitin ligase complex that is a part of the ubiquitin-proteasome protein degradation pathway. ARF proteins have DNA binding domains and can bind promoter regions of genes and activate or repress gene expression. Aux/IAA proteins can bind ARF proteins sitting on gene promoters and prevent them from doing their job. TIR1 proteins are F-box proteins that have three different domains giving them the ability to bind to three different ligands: an SCF^{TIR1} ubiquitin ligase complex (using the F-box domain), auxin (so TIR1 proteins are auxin receptors), and Aux/IAA proteins (via a degron domain). Upon binding of auxin, a TIR1 protein's degron domain has increased affinity for Aux/IAA repressor proteins, which when bound to TIR1 and its SCF complex undergo ubiquitination and subsequent degradation by a proteasome. The degradation of Aux/IAA proteins frees ARF proteins to activate or repress genes at whose promoters they are bound.

Within a plant, elaboration of the Aux/IAA repressor pathway takes place via diversification of the TIR1, ARF, and Aux/IAA protein families. Each family may contain many similar-acting proteins, differing in qualities such as degree of affinity for partner proteins, amount of activation or repression of target gene transcription, or domains of expression (e.g. different plant tissues might express different members of the family, or different environmental stresses might activate expression of different members). Such elaboration permits the plant to use auxin in a variety of ways depending on the needs of the tissue and plant.

Other Auxin Regulatory Pathways

Another protein, auxin-binding protein 1 (ABP1), is a putative receptor for a different signaling pathway, but its role is as yet unclear. Electrophysiological experiments with protoplasts and anti-ABP1 antibodies suggest ABP1 may have a function at the plasma membrane, and cells can possibly use ABP1 proteins to respond to auxin through means faster and independent of gene expression.

On a Cellular Level

On the cellular level, auxin is essential for cell growth, affecting both cell division and cellular expansion. Auxin concentration level, together with other local factors, contributes to cell differentiation and specification of the cell fate.

Depending on the specific tissue, auxin may promote axial elongation (as in shoots), lateral expansion (as in root swelling), or isodiametric expansion (as in fruit growth). In some cases (coleoptile growth), auxin-promoted cellular expansion occurs in the absence of cell division. In other cases, auxin-promoted cell division and cell expansion may be closely sequenced within the same tissue (root initiation, fruit growth). In a living plant, auxins and other plant hormones nearly always appear to interact to determine patterns of plant development.

Organ Patterns

Growth and division of plant cells together result in growth of tissue, and specific tissue growth contributes to the development of plant organs.

Growth of cells contributes to the plant's size, unevenly localized growth produces bending, turning and directionalization of organs- for example, stems turning toward light sources (phototropism), roots growing in response to gravity (gravitropism), and other tropisms originated because cells on one side grow faster than the cells on the other side of the organ. So, precise control of auxin distribution between different cells has paramount importance to the resulting form of plant growth and organization.

Auxin Transport and the Uneven Distribution of Auxin

To cause growth in the required domains, auxins must of necessity be active preferentially in them. Local auxin maxima can be formed by active biosynthesis in certain cells of tissues, for example via tryptophan-dependent pathways, but auxins are not synthesized in all cells (even if cells retain the potential ability to do so, only under specific conditions will auxin synthesis be activated in them). For that purpose, auxins have to be not only translocated toward those sites where they are needed, but also they must have an established mechanism to detect those sites. For that purpose, auxins have to be translocated toward those sites where they are needed. Translocation is driven throughout the plant body, primarily from peaks of shoots to peaks of roots (from up to down).

For long distances, relocation occurs via the stream of fluid in phloem vessels, but, for short-distance transport, a unique system of coordinated polar transport directly from cell to cell is exploited. This short-distance, active transport exhibits some morphogenetic properties.

This process, polar auxin transport, is directional, very strictly regulated, and based in uneven distribution of auxin efflux carriers on the plasma membrane, which send auxins in the proper direction. While PIN-FORMED (PIN) proteins are vital in transporting auxin in a polar manner, the family of AUXIN1/LIKE-AUX1 (AUX/LAX) genes encodes for non-polar auxin influx carriers.

The regulation of PIN protein localisation in a cell determines the direction of auxin transport from cell, and concentrated effort of many cells creates peaks of auxin, or auxin maxima (regions having cells with higher auxin – a maximum). Proper and timely auxin maxima within developing roots and shoots are necessary to organise the development of the organ. Surrounding auxin maxima are cells with low auxin troughs, or auxin minima. For example, in the *Arabidopsis* fruit, auxin minima have been shown to be important for its tissue development.

Organization of the Plant

As auxins contribute to organ shaping, they are also fundamentally required for proper development of the plant itself. Without hormonal regulation and organization, plants would be merely proliferating heaps of similar cells. Auxin employment begins in the embryo of the plant, where directional distribution of auxin ushers in subsequent growth and development of primary growth poles, then forms buds of future organs. Next, it helps to coordinate proper development of the arising organs, such as roots, cotyledons and leaves and mediates long distance signals between them, contributing so to the overall architecture of the plant. Throughout the plant's life, auxin helps the plant maintain the polarity of growth, and actually "recognize" where it has its branches (or any organ) connected.

An important principle of plant organization based upon auxin distribution is apical dominance, which means the auxin produced by the apical bud (or growing tip) diffuses (and is transported) downwards and inhibits the development of ulterior lateral bud growth, which would otherwise compete with the apical tip for light and nutrients. Removing the apical tip and its suppressively acting auxin allows the lower dormant lateral buds to develop, and the buds between the leaf stalk and stem produce new shoots which compete to become the lead growth. The process is actually quite complex, because auxin transported downwards from the lead shoot tip has to interact with several other plant hormones (such as strigolactones or cytokinins) in the process on various positions along the growth axis in plant body to achieve this phenomenon. This plant behavior is used in pruning by horticulturists.

Finally, the sum of auxin arriving from stems to roots influences the degree of root growth. If shoot tips are removed, the plant does not react just by outgrowth of lateral buds — which are supposed to replace to original lead. It also follows that smaller amount of auxin arriving to the roots results in slower growth of roots and the nutrients are subsequently in higher degree invested in the upper part of the plant, which hence starts to grow faster.

Effects

Auxin participates in phototropism, geotropism, hydrotropism and other developmental changes. The uneven distribution of auxin, due to environmental cues, such as unidirectional light or gravity force, results in uneven plant tissue growth, and generally, auxin governs the form and shape of plant body, direction and strength of growth of all organs, and their mutual interaction.

Auxin stimulates cell elongation by stimulating wall-loosening factors, such as elastins, to loosen cell walls. The effect is stronger if gibberellins are also present. Auxin also stimulates cell division if cytokinins are present. When auxin and cytokinin are applied to callus, rooting can be generated if the auxin concentration is higher than cytokinin concentration. Xylem tissues can be generated when the auxin concentration is equal to the cytokinins.

Auxin also induces sugar and mineral accumulation at the site of application.

A healthy Arabidopsis thaliana plant (left) next to an auxin signal-transduction mutant

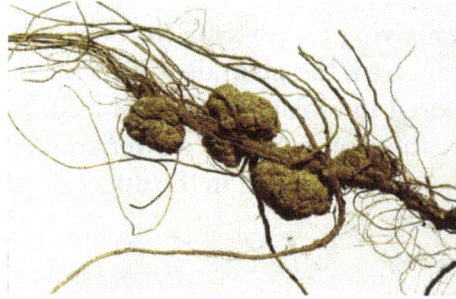

Crown galls are caused by Agrobacterium tumefaciens bacteria; they produce and excrete auxin and cytokinin, which interfere with normal cell division and cause tumors.

Wound Response

Auxin induces the formation and organization of phloem and xylem. When the plant is wounded, the auxin may induce the cell differentiation and regeneration of the vascular tissues.

Root Growth and Development

Auxins promote root initiation. Auxin induces both growth of pre-existing roots and adventitious root formation, i.e., branching of the roots. As more native auxin is transported down the stem to the roots, the overall development of the roots is stimulated. If the source of auxin is removed, such as by trimming the tips of stems, the roots are less stimulated accordingly, and growth of stem is supported instead.

In horticulture, auxins, especially NAA and IBA, are commonly applied to stimulate root initiation when rooting cuttings of plants. However, high concentrations of auxin inhibit root elongation and instead enhance adventitious root formation. Removal of the root tip can lead to inhibition of secondary root formation.

Apical Dominance

Auxin induces shoot apical dominance; the axillary buds are inhibited by auxin, as a high concentration of auxin directly stimulates ethylene synthesis in axillary buds, causing inhibition of their growth and potentiation of apical dominance. When the apex of the plant is removed, the inhibitory effect is removed and the growth of lateral buds is enhanced. Auxin is sent to the part of the plant facing away from the light, where it promotes cell elongation, thus causing the plant to bend towards the light.

Fruit Growth and Development

Auxin is required for fruit growth and development and delays fruit senescence. When seeds are removed from strawberries, fruit growth is stopped; exogenous auxin stimulates the growth in fruits with seeds removed. For fruit with unfertilized seeds, exogenous auxin results in parthenocarpy ("virgin-fruit" growth).

Fruits form abnormal morphologies when auxin transport is disturbed. In *Arabidopsis* fruits, auxin controls the release of seeds from the fruit (pod). The valve margins are a specialised tissue in pods that regulates when pod will open (dehiscence). Auxin must be removed from the valve margin cells to allow the valve margins to form. This process requires modification of the auxin transporters (PIN proteins).

Flowering

Auxin plays also a minor role in the initiation of flowering and development of reproductive organs. In low concentrations, it can delay the senescence of flowers. A number of plant mutants have been described that affect flowering and have deficiencies in either auxin synthesis or transport. In maize, one example is bif2 barren inflorescence2.

Ethylene Biosynthesis

In low concentrations, auxin can inhibit ethylene formation and transport of precursor in plants; however, high concentrations can induce the synthesis of ethylene. Therefore, the high concentration can induce femaleness of flowers in some species.

Auxin inhibits abscission prior to formation of abscission layer, and thus inhibits senescence of leaves.

Synthetic Auxins

In the course of research on auxin biology, many compounds with noticeable auxin activity were synthesized. Many of them had been found to have economical potential for man-controlled growth and development of plants in agronomy. Synthetic auxins include the following compounds:

- Gallery of synthetic auxins

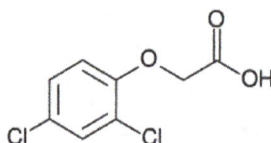

2,4-Dichlorophenoxyacetic acid (2,4-D); active herbicide and main auxin in laboratory use

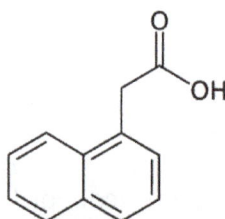

α-Naphthalene acetic acid (α-NAA); often part of commercial rooting powders

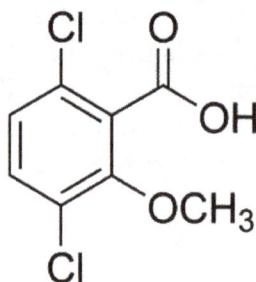

2-Methoxy-3,6-dichlorobenzoic acid (dicamba); active herbicide

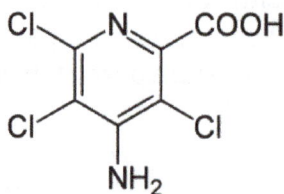

4-Amino-3,5,6-trichloropicolinic acid (tordon or picloram); active herbicide

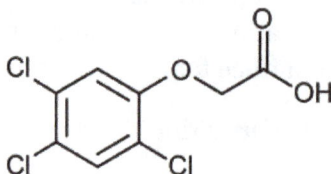

2,4,5-Trichlorophenoxyacetic acid (2,4,5-T)

Auxins are toxic to plants in large concentrations; they are most toxic to dicots and less so to monocots. Because of this property, synthetic auxin herbicides, including 2,4-D and 2,4,5-T, have been developed and used for weed control.

However, some exogenously synthesized auxins, especially 1-naphthaleneacetic acid (NAA) and indole-3-butyric acid (IBA), are also commonly applied to stimulate root growth when taking cuttings of plants or for different agricultural purposes such as the prevention of fruit drop in orchards.

Used in high doses, auxin stimulates the production of ethylene. Excess ethylene (also native plant hormone) can inhibit elongation growth, cause leaves to fall (abscission), and even kill the plant. Some synthetic auxins, such as 2,4-D and 2,4,5-trichlorophenoxyacetic acid (2,4,5-T) were marketed also as herbicides. Dicots, such as dandelions, are much more susceptible to auxins than monocots, such as grasses and cereal crops. So, these synthetic auxins are valuable as synthetic herbicides. 2,4-D was the first widely used herbicide, and it is still so. 2,4-D was first commercialized by the Sherwin-Williams company, and saw use in the late 1940s. It is easy and inexpensive to manufacture.

Herbicide manufacture

The defoliant Agent Orange, used extensively by British forces in the Malayan Emergency and American forces in the Vietnam War, was a mix of 2,4-D and 2,4,5-T. The compound 2,4-D is still in use and is thought to be safe, but 2,4,5-T was more or less banned by the U.S. Environmental Protection Agency in 1979. The dioxin TCDD is an unavoidable contaminant produced in the manufacture of 2,4,5-T. As a result of the integral dioxin contamination, 2,4,5-T has been implicated in leukemia, miscarriages, birth defects, liver damage, and other diseases.

Cytokinin

Cytokinins (CK) are a class of plant growth substances (phytohormones) that promote cell division, or cytokinesis, in plant roots and shoots. They are involved primarily in cell growth and differentiation, but also affect apical dominance, axillary bud growth, and leaf senescence. Folke Skoog discovered their effects using coconut milk in the 1940s at the University of Wisconsin–Madison.

There are two types of cytokinins: adenine-type cytokinins represented by kinetin, zeatin, and 6-benzylaminopurine, and phenylurea-type cytokinins like diphenylurea and thidiazuron (TDZ). Most adenine-type cytokinins are synthesized in roots. Cambium and other actively dividing tissues also synthesize cytokinins. No phenylurea cytokinins have been found in plants. Cytokinins participate in local and long-distance signalling, with the same transport mechanism as purines and nucleosides. Typically, cytokinins are transported in the xylem.

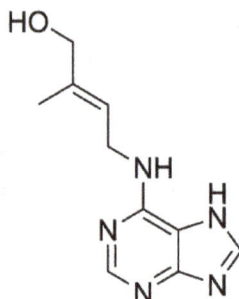

The cytokinin zeatin is named after the genus of corn, *Zea*, in which it was discovered.

Cytokinins act in concert with auxin, another plant growth hormone. The two are complementary, having generally opposite effects.

Mode of Action

The ratio of auxin to cytokinin plays an important role in the effect of cytokinin on plant growth. Cytokinin alone has no effect on parenchyma cells. When cultured with auxin but no cytokinin, they grow large but do not divide. When cytokinin is added, the cells expand and differentiate. When cytokinin and auxin are present in equal levels, the parenchyma cells form an undifferentiated callus. More cytokinin induces growth of shoot buds, while more auxin induces root formation.

Cytokinins are involved in many plant processes, including cell division and shoot and root morphogenesis. They are known to regulate axillary bud growth and apical dominance. The "direct inhibition hypothesis" posits that these effects result from the cytokinin to auxin ratio. This theory states that auxin from apical buds travels down shoots to inhibit axiliary bud growth. This promotes shoot growth, and restricts lateral branching. Cytokinin moves from the roots into the shoots, eventually signaling lateral bud growth. Simple experiments support this theory. When the apical bud is removed, the axillary buds are uninhibited, lateral growth increases, and plants become bushier. Applying auxin to the cut stem again inhibits lateral dominance.

While cytokinin action in vascular plants is described as pleiotropic, this class of plant hormones specifically induces the transition from apical growth to growth via a three-faced apical cell in moss protonema. This bud induction can be pinpointed to differentiation of a specific single cell, and thus is a very specific effect of cytokinin.

Cytokinins have been shown to slow aging of plant organs by preventing protein breakdown, activating protein synthesis, and assembling nutrients from nearby tissues. A study that regulated leaf senescence in tobacco leaves found that wild-type leaves yellowed while transgenic leaves remained mostly green. It was hypothesized that cytokinin may affect enzymes that regulate protein synthesis and degradation.

Cytokinin signaling in plants is mediated by a two-component phosphorelay. This pathway is initiated by cytokinin binding to a histidine kinase receptor in the endoplasmic reticulum membrane. This results in the autophosphorylation of the receptor, with the phosphate then being transferred to a phosphotransfer protein. The phosphotransfer proteins can then phosphorylate the type-B response regulators (RR) which are a family of transcriptions factors. The phosphorylated, and thus activated, type-B RRs regulate the transcription of numerous genes, including the type-A *RR*s. The type-A RRs negatively regulate the pathway.

Biosynthesis

Adenosine phosphate-isopentenyltransferase (IPT) catalyses the first reaction in the biosynthesis of isoprene cytokinins. It may use ATP, ADP, or AMP as substrates and may use dimethylallyl pyrophosphate (DMAPP) or hydroxymethylbutenyl pyrophosphate (HMBPP) as prenyl donors. This reaction is the rate-limiting step in cytokinin biosynthesis. DMADP and HMBDP used in cytokinin biosynthesis are produced by the methylerythritol phosphate pathway (MEP).

Cytokinins can also be produced by recycled tRNAs in plants and bacteria. tRNAs with anticodons that start with a uridine and carrying an already-prenylated adenosine adjacent to the anticodon release on degradation the adenosine as a cytokinin. The prenylation of these adenines is carried out by tRNA-isopentenyltransferase.

Auxin is known to regulate the biosynthesis of cytokinin.

Uses

Because cytokinin promotes plant cell division and growth, produce farmers use it to increase crops. One study found that applying cytokinin to cotton seedlings led to a 5–10% yield increase under drought conditions.

Cytokinins have recently been found to play a role in plant pathogenesis. For example, cytokinins have been described to induce resistance against *Pseudomonas syringae* in *Arabidopsis thaliana* and *Nicotiana tabacum*. Also in context of biological control of plant diseases cytokinins seem to have potential functions. Production of cytokinins by *Pseudomonas fluorescens* G20-18 has been identified as a key determinant to efficiently control the infection of *A. thaliana* with *P. syringae*.

Ethylene

Ethylene (IUPAC name: ethene) is a hydrocarbon which has the formula C_2H_4 or $H_2C=CH_2$. It is a colorless flammable gas with a faint "sweet and musky" odour when pure. It is the simplest alkene (a hydrocarbon with carbon-carbon double bonds), and the second simplest unsaturated hydrocarbon after acetylene (C2H2).

Ethylene is widely used in the chemical industry, and its worldwide production (over 150 million tonnes in 2016) exceeds that of any other organic compound. Much of this production goes toward polyethylene, a widely used plastic containing polymer chains of ethylene units in various chain lengths. Ethylene is also an important natural plant hormone, used in agriculture to force the ripening of fruits. Ethylene's hydrate is ethyl alcohol.

Structure and Properties

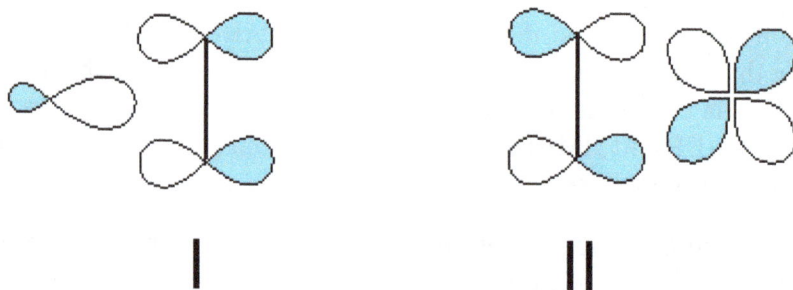

Orbital description of bonding between ethylene and a transition metal.

This hydrocarbon has four hydrogen atoms bound to a pair of carbon atoms that are connected by a double bond. All six atoms that comprise ethylene are coplanar. The H-C-H angle is 117.4°, close to the 120° for ideal sp^2 hybridized carbon. The molecule is also relatively rigid: rotation about the C-C bond is a high energy process that requires breaking the π-bond.

The π-bond in the ethylene molecule is responsible for its useful reactivity. The double bond is a region of high electron density, thus it is susceptible to attack by electrophiles. Many reactions of ethylene are catalyzed by transition metals, which bind transiently to the ethylene using both the π and π* orbitals.

Being a simple molecule, ethylene is spectroscopically simple. Its UV-vis spectrum is still used as a test of theoretical methods.

Uses

Major industrial reactions of ethylene include in order of scale: 1) polymerization, 2) oxidation, 3) halogenation and hydrohalogenation, 4) alkylation, 5) hydration, 6) oligomerization, and 7) hydroformylation. In the United States and Europe, approximately 90% of ethylene is used to produce ethylene oxide, ethylene dichloride, ethylbenzene and polyethylene. Most of the reactions with ethylene are electrophilic addition.

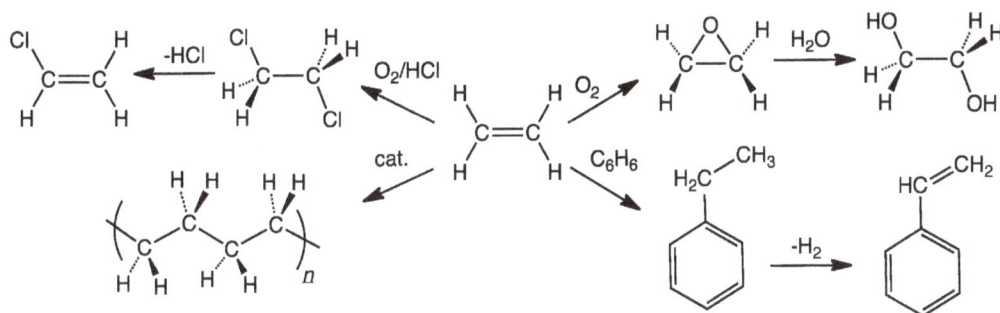

Main industrial uses of ethylene. Clockwise from the upper right: its conversions to ethylene oxide, precursor to ethylene glycol; to ethylbenzene, precursor to styrene; to various kinds of polyethylene; to ethylene dichloride, precursor to vinyl chloride.

Polymerization

Polyethylene consumes more than half of the world's ethylene supply. Polyethylene, also called *polythene*, is the world's most widely used plastic. It is primarily used to make films in packaging, carrier bags and trash liners. Linear alpha-olefins, produced by oligomerization (formation of short polymers) are used as precursors, detergents, plasticisers, synthetic lubricants, additives, and also as co-monomers in the production of polyethylenes.

Oxidation

Ethylene is oxidized to produce ethylene oxide, a key raw material in the production of surfactants and detergents by ethoxylation. Ethylene oxide is also hydrolyzed to produce ethylene glycol, widely used as an automotive antifreeze as well as higher molecular weight glycols, glycol ethers and polyethylene terephthalate.

Ethylene undergoes oxidation by palladium to give acetaldehyde. This conversion remains a major industrial process (10M kg/y). The process proceeds via the initial complexation of ethylene to a Pd(II) center.

Halogenation and Hydrohalogenation

Major intermediates from the halogenation and hydrohalogenation of ethylene include ethylene dichloride, ethyl chloride and ethylene dibromide. The addition of chlorine entails "oxychlorination," i.e. chlorine itself is not used. Some products derived from this group are polyvinyl chloride, trichloroethylene, perchloroethylene, methyl chloroform, polyvinylidene chloride and copolymers, and ethyl bromide.

Alkylation

Major chemical intermediates from the alkylation with ethylene is ethylbenzene, precursor to styrene. Styrene is used principally in polystyrene for packaging and insulation, as well as in styrene-butadiene rubber for tires and footwear. On a smaller scale, ethyltoluene, ethylanilines, 1,4-hexadiene, and aluminium alkyls. Products of these intermediates include polystyrene, unsaturated polyesters and ethylene-propylene ter-polymers.

Oxo Reaction

The hydroformylation (oxo reaction) of ethylene results in propionaldehyde, a precursor to propionic acid and n-propyl alcohol.

Hydration

Ethylene has long represented the major nonfermentative precursor to ethanol. The original method entailed its conversion to diethyl sulfate, followed by hydrolysis. The main method practiced since the mid-1990s is the direct hydration of ethylene catalyzed by solid acid catalysts:

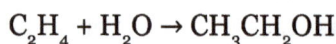

$$C_2H_4 + H_2O \rightarrow CH_3CH_2OH$$

Dimerization to n-Butenes

Ethylene can be dimerized to n-butenes using processes licensed by Lummus or IFP. The Lummus process produces mixed n-butenes (primarily 2-butenes) while the IFP process produces 1-butene.

Niche Uses

An example of a niche use is as an anesthetic agent (in an 85% ethylene/15% oxygen ratio). It can also be used to hasten fruit ripening, as well as a welding gas.

Production

Global ethylene production was 107 million tonnes in 2005, 109 million tonnes in 2006, 138 million tonnes in 2010 and 141 million tonnes in 2011. By 2013 ethylene was produced by at least 117 companies in 32 countries. To meet the ever increasing demand for ethylene, sharp increases in production facilities are added globally, particularly in the Mideast and in China.

Industrial Process

Ethylene is produced in the petrochemical industry by steam cracking. In this process, gaseous or light liquid hydrocarbons are heated to 750–950 °C, inducing numerous free radical reactions followed by immediate quench to stop these reactions. This process converts large hydrocarbons into smaller ones and introduces unsaturation. Ethylene is separated from the resulting mixture by repeated compression and distillation. In a related process used in oil refineries, high molecular weight hydrocarbons are cracked over zeolite catalysts. Heavier feedstocks, such as naphtha and gas oils require at least two "quench towers" downstream of the cracking furnaces to recirculate pyrolysis-derived gasoline and process water. When cracking a mixture of ethane and propane, only one water quench tower is required.

The areas of an ethylene plant are:

1. steam cracking furnaces:

2. primary and secondary heat recovery with quench;

3. a dilution steam recycle system between the furnaces and the quench system;

4. primary compression of the cracked gas (3 stages of compression);

5. hydrogen sulfide and carbon dioxide removal (acid gas removal);

6. secondary compression (1 or 2 stages);

7. drying of the cracked gas;

8. cryogenic treatment;

9. all of the cold cracked gas stream goes to the demethanizer tower. The overhead stream from the demethanizer tower consists of all the hydrogen and methane that was in the cracked gas stream. Cryogenically (−250 °F (−157 °C)) treating this overhead stream separates hydrogen from methane. Methane recovery is critical to the economical operation of an ethylene plant.

10. the bottom stream from the demethanizer tower goes to the deethanizer tower. The overhead stream from the deethanizer tower consists of all the C2's that were in the cracked gas stream. The C2 stream contains acetylene, which is explosive above 200 kPa (29 psi). If the partial pressure of acetylene is expected to exceed these values, the C2 stream is partially hydrogenated. The C2's then proceed to a C2 splitter. The product ethylene is taken from the overhead of the tower and the ethane coming from the bottom of the splitter is recycled to the furnaces to be cracked again;

11. the bottom stream from the de-ethanizer tower goes to the depropanizer tower. The overhead stream from the depropanizer tower consists of all the C3's that were in the cracked gas stream. Before feeding the C3's to the C3 splitter, the stream is hydrogenated to convert the methylacetylene and propadiene (allene) mix. This stream is then sent to the C3 splitter. The overhead stream from the C3 splitter is product propylene and the bottom stream is propane which is sent back to the furnaces for cracking or used as fuel.

12. The bottom stream from the depropanizer tower is fed to the debutanizer tower. The overhead stream from the debutanizer is all of the C4's that were in the cracked gas stream. The bottom stream from the debutanizer (light pyrolysis gasoline) consists of everything in the cracked gas stream that is C5 or heavier.

Since ethylene production is energy intensive, much effort has been dedicated to recovering heat from the gas leaving the furnaces. Most of the energy recovered from the cracked gas is used to make high pressure (1200 psig) steam. This steam is in turn used to drive the turbines for compressing cracked gas, the propylene refrigeration compressor, and the ethylene refrigeration compressor. An ethylene plant, once running, does not need to import steam to drive its steam turbines. A typical world scale ethylene plant (about 1.5 billion pounds of ethylene per year) uses a 45,000 horsepower (34,000 kW) cracked gas compressor, a 30,000 hp (22,000 kW) propylene compressor, and a 15,000 hp (11,000 kW) ethylene compressor.

Laboratory Synthesis

Although of great value industrially, ethylene is rarely used in the laboratory and is ordinarily purchased. It can be produced via dehydration of ethanol with sulfuric acid or in the gas phase with aluminium oxide.

Ethylene as a Plant Hormone

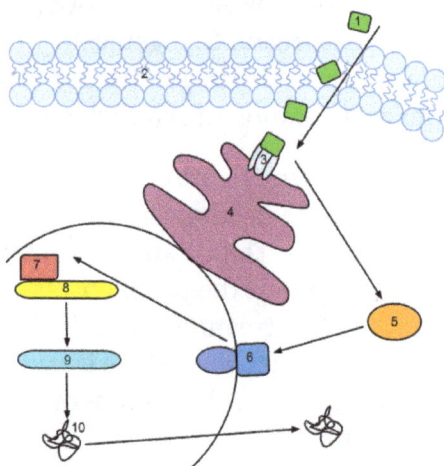

An ethylene signal transduction pathway. Ethylene permeates the membrane and binds to a receptor on the endoplasmic reticulum. The receptor releases the repressed EIN2. This then activates a signal transduction pathway which activates a regulatory genes that eventually trigger an Ethylene response. The activated DNA is transcribed into mRNA which is then translated into a functional enzyme that is used for ethylene biosynthesis.

Ethylene serves as a hormone in plants. It acts at trace levels throughout the life of the plant by stimulating or regulating the ripening of fruit, the opening of flowers, and the abscission (or shedding) of leaves. Commercial ripening rooms use "catalytic generators" to make ethylene gas from a liquid supply of ethanol. Typically, a gassing level of 500 to 2,000 ppm is used, for 24 to 48 hours. Care must be taken to control carbon dioxide levels in ripening rooms when gassing, as high temperature ripening (20 °C; 68 °F) has been seen to produce CO_2 levels of 10% in 24 hours.

History of Ethylene in Plant Biology

Ethylene has been used since the ancient Egyptians, who would gash figs in order to stimulate ripening (wounding stimulates ethylene production by plant tissues). The ancient Chinese would burn incense in closed rooms to enhance the ripening of pears. In 1864, it was discovered that gas leaks from street lights led to stunting of growth, twisting of plants, and abnormal thickening of stems. In 1901, a Russian scientist named Dimitry Neljubow showed that the active component was ethylene. Sarah Doubt discovered that ethylene stimulated abscission in 1917. It was not until 1934 that Gane reported that plants synthesize ethylene. In 1935, Crocker proposed that ethylene was the plant hormone responsible for fruit ripening as well as senescence of vegetative tissues.

Ethylene Biosynthesis in Plants

The Yang cycle

Ethylene is produced from essentially all parts of higher plants, including leaves, stems, roots, flowers, fruits, tubers, and seeds. Ethylene production is regulated by a variety of developmental and environmental factors. During the life of the plant, ethylene production is induced during certain stages of growth such as germination, ripening of fruits, abscission of leaves, and senescence of flowers. Ethylene production can also be induced by a variety of external aspects such as mechanical wounding, environmental stresses, and certain chemicals including auxin and other regulators. The pathway for ethylene biosynthesis is named the Yang cycle after the scientist Shang Fa Yang who made key contributions to elucidating this pathway.

Ethylene is biosynthesized from the amino acid methionine to S-adenosyl-L-methionine (SAM, also called Adomet) by the enzyme Met Adenosyltransferase. SAM is then converted to 1-aminocyclopropane-1-carboxylic acid (ACC) by the enzyme ACC synthase (ACS). The activity of ACS determines the rate of ethylene production, therefore regulation of this enzyme is key for the ethylene biosynthesis. The final step requires oxygen and involves the action of the enzyme ACC-oxidase (ACO), formerly known as the ethylene forming enzyme (EFE). Ethylene biosynthesis can be induced by endogenous or exogenous ethylene. ACC synthesis increases with high levels of auxins, especially indole acetic acid (IAA) and cytokinins.

Ethylene Perception in Plants

Ethylene is perceived by a family of five transmembrane protein dimers such as the ETR1 protein in *Arabidopsis*. The genes encoding ethylene receptors have been cloned in the reference plant *Arabidopsis thaliana* and many other plants. Ethylene receptors are encoded by multiple genes in plant genomes. Dominant missense mutations in any of the gene family, which comprises five receptors in *Arabidopsis* and at least six in tomato, can confer insensitivity to ethylene. Loss-of-function mutations in multiple members of the ethylene-receptor family result in a plant that exhibits constitutive ethylene responses. DNA sequences for ethylene receptors have also been identified in many other plant species and an ethylene binding protein has even been identified in Cyanobacteria.

Environmental and Biological Triggers of Ethylene

Environmental cues such as flooding, drought, chilling, wounding, and pathogen attack can induce ethylene formation in plants. In flooding, roots suffer from lack of oxygen, or anoxia, which leads to the synthesis of 1-aminocyclopropane-1-carboxylic acid (ACC). ACC is transported upwards in the plant and then oxidized in leaves. The ethylene produced causes nastic movements (epinasty) of the leaves, perhaps helping the plant to lose water.

List of Plant Responses to Ethylene

- Seedling triple response, thickening and shortening of hypocotyl with pronounced apical hook.

- Stimulation of *Arabidopsis* hypocotyl elongation

- In pollination, when the pollen reaches the stigma, the precursor of the ethene, ACC, is secreted to the petal, the ACC releases ethylene with ACC oxidase.

- Stimulates leaf and flower senescence

- Stimulates senescence of mature xylem cells in preparation for plant use

- Induces leaf abscission

- Induces seed germination

- Induces root hair growth — increasing the efficiency of water and mineral absorption

- Induces the growth of adventitious roots during flooding

- Stimulates epinasty — leaf petiole grows out, leaf hangs down and curls into itself

- Stimulates fruit ripening

- Induces a climacteric rise in respiration in some fruit which causes a release of additional ethylene.

- Affects gravitropism

- Stimulates nutational bending

- Inhibits stem growth and stimulates stem and cell broadening and lateral branch growth outside of seedling stage

- Interference with auxin transport (with high auxin concentrations)

- Inhibits shoot growth and stomatal closing except in some water plants or habitually flooded ones such as some rice varieties, where the opposite occurs (conserving CO_2 and O_2)

- Induces flowering in pineapples

- Inhibits short day induced flower initiation in *Pharbitus nil* and *Chrysanthemum morifolium*

Commercial Issues

Ethylene shortens the shelf life of many fruits by hastening fruit ripening and floral senescence. Ethylene will shorten the shelf life of cut flowers and potted plants by accelerating floral senescence and floral abscission. Flowers and plants which are subjected to stress during shipping, handling, or storage produce ethylene causing a significant reduction in floral display. Flowers affected by ethylene include carnation, geranium, petunia, rose, and many others.

Ethylene can cause significant economic losses for florists, markets, suppliers, and growers. Researchers have developed several ways to inhibit ethylene, including inhibiting ethylene synthesis and inhibiting ethylene perception. Aminoethoxyvinylglycine (AVG), Aminooxyacetic acid (AOA), and silver salts are ethylene inhibitors. Inhibiting ethylene synthesis is less effective for reducing post-harvest losses since ethylene from other sources can still have an effect. By inhibiting ethylene perception, fruits, plants and flowers don't respond to ethylene produced endogenously or from exogenous sources. Inhibitors of ethylene perception include compounds that have a similar shape to ethylene, but do not elicit the ethylene response. One example of an ethylene perception inhibitor is 1-methylcyclopropene (1-MCP).

Commercial growers of bromeliads, including pineapple plants, use ethylene to induce flowering. Plants can be induced to flower either by treatment with the gas in a chamber, or by placing a banana peel next to the plant in an enclosed area.

Chrysanthemum flowering is delayed by ethylene gas and growers have found that carbon dioxide 'burners' and the exhaust fumes from inefficient glasshouse heaters can raise the ethylene concentration to 0.05 ppmv causing delay in flowering of commercial crops.

Ligand

Ethylene is a ligand in organometallic chemistry. One of the first organometallic compounds, Zeise's salt is a complex of ethylene. Useful reagents containing ethylene include $Pt(PPh_3)_2(C_2H_4)$ and $Rh_2Cl_2(C_2H_4)_4$. The Rh-catalysed hydroformylation of ethylene is conducted on industrial scale to provide propionaldehyde.

History

Some geologists and scholars believe that the famous Greek Oracle at Delphi (the Pythia) went into her trance-like state as an effect of ethylene rising from ground faults.

Ethylene appears to have been discovered by Johann Joachim Becher, who obtained it by heating ethanol with sulfuric acid; he mentioned the gas in his *Physica Subterranea* (1669). Joseph Priestley also mentions the gas in his *Experiments and observations relating to the various branches of natural philosophy: with a continuation of the observations on air* (1779), where he reports that Jan Ingenhousz saw ethylene synthesized in the same way by a Mr. Enée in Amsterdam in 1777 and that Ingenhousz subsequently produced the gas himself. The properties of ethylene were studied in 1795 by four Dutch chemists, Johann Rudolph Deimann, Adrien Paets van Troostwyck, Anthoni Lauwerenburgh and Nicolas Bondt, who found that it differed from hydrogen gas and that it contained both carbon and hydrogen. This group also discovered that ethylene could be combined with chlorine to produce the *oil of the Dutch chemists*, 1,2-dichloroethane; this discovery gave ethylene the name used for it at that time, *olefiant gas* (oil-making gas.)

In the mid-19th century, the suffix *-ene* was widely used to refer to a molecule or part thereof that contained one fewer hydrogen atoms than the molecule being modified. Thus, *ethylene* (C2H4) was the "daughter of ethyl" (C2H5). The name ethylene was used in this sense as early as 1852.

In 1866, the German chemist August Wilhelm von Hofmann proposed a system of hydrocarbon nomenclature in which the suffixes -ane, -ene, -ine, -one, and -une were used to denote the hydrocarbons with 0, 2, 4, 6, and 8 fewer hydrogens than their parent alkane. In this system, ethylene became *ethene*. Hofmann's system eventually became the basis for the Geneva nomenclature approved by the International Congress of Chemists in 1892, which remains at the core of the IUPAC nomenclature. However, by that time, the name ethylene was deeply entrenched, and it remains in wide use today, especially in the chemical industry.

Following experimentation by Luckhardt, Crocker, and Carter at the University of Chicago, ethylene was used as an anesthetic It remained in use through the 1940s use even while chloroform was being phased out. Its pungent odor and its explosive nature limit its use today.

Nomenclature

The 1979 IUPAC nomenclature rules made an exception for retaining the non-systematic name ethylene, however, this decision was reversed in the 1993 rules so the IUPAC name is now *ethene*.

Safety

Like all hydrocarbons, ethylene is an asphyxiant and combustible. It is listed as an IARC class 3 carcinogen.

Gibberellin

Gibberellins (GAs) are plant hormones that regulate growth and influence various developmental processes, including stem elongation, germination, dormancy, flowering, sex expression, enzyme induction, and leaf and fruit senescence.

Gibberellin was first recognized in 1926 by a Japanese scientist, Eiichi Kurosawa, studying *bakanae*, the "foolish seedling" disease in rice. It was first isolated in 1935 by Teijiro Yabuta and Sumuki, from fungal strain (*Gibberella fujikuroi*) provided by Kurosawa. Yabuta named the isolate as gibberellin.

Interest in gibberellins outside Japan began after World War II. In the United States, the first research was undertaken by a unit at Camp Detrick in Maryland, via studying seedlings of the bean *Vicia faba*. In the United Kingdom, work on isolating new types of gibberellin was undertaken at Imperial Chemical Industries. Interest in gibberellins spread around the world as the potential for its use on various commercially important plants became more obvious. For example, research that started at the University of California, Davis in the mid-1960s led to its commercial use on Thompson seedless table grapes throughout California by 1962. A known antagonist to gibberellin is paclobutrazol (PBZ), which in turn inhibits growth and induces early fruitset as well as seedset.

Chemistry

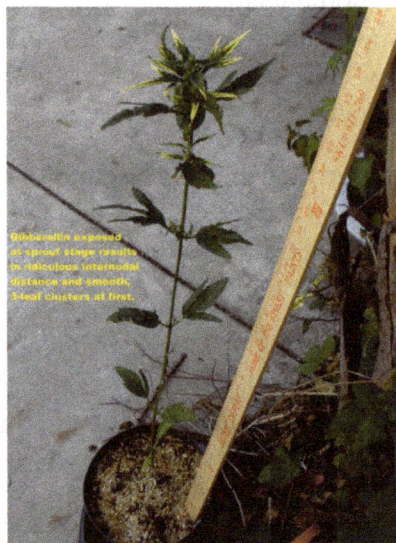

Effect of gibberellic acid on cannabis sprout

All known gibberellins are diterpenoid acids that are synthesized by the terpenoid pathway in plastids and then modified in the endoplasmic reticulum and cytosol until they reach their biologically-active form. All gibberellins are derived via the *ent*-gibberellane skeleton, but are synthesised via *ent*-kaurene. The gibberellins are named GA1 through GAn in order of discovery. Gibberellic acid, which was the first gibberellin to be structurally characterized, is GA3.

As of 2003, there were 126 GAs identified from plants, fungi, and bacteria.

Gibberellins are tetracyclic diterpene acids. There are two classes based on the presence of either 19 or 20 carbons. The 19-carbon gibberellins, such as gibberellic acid, have lost carbon 20 and, in place, possess a five-member lactone bridge that links carbons 4 and 10. The 19-carbon forms are, in general, the biologically active forms of gibberellins. Hydroxylation also has a great effect on the biological activity of the gibberellin. In general, the most biologically active compounds are dihydroxylated gibberellins, which possess hydroxyl groups on both carbon 3 and carbon 13. Gibberellic acid is a dihydroxylated gibberellin.

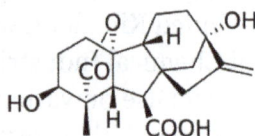

Gibberellin A1 (GA1)

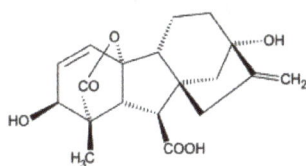

Gibberellic acid (GA3)

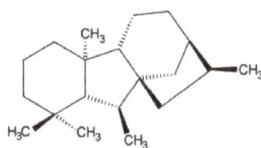

ent-Gibberellane

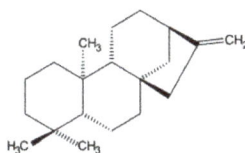

ent-Kaurene

Bioactive GAs

The bioactive GAs are GA1, GA3, GA4, and GA7. There are three common structural traits between these GAs: 1) a hydroxyl group on C-3β, 2) a carboxyl group on C-6, and 3) a lactone between C-4 and C-10. The 3β-hydroxyl group can be exchanged for other functional groups at C-2 and/or C-3 positions. GA5 and GA6 are examples of bioactive GAs that do not have a hydroxyl group on C-3β. The presence of GA1 in various plant species suggests that it is a common bioactive GA.

Biological Function

Gibberellins are involved in the natural process of breaking dormancy and other aspects of germination. Before the photosynthetic apparatus develops sufficiently in the early stages of germination, the stored energy reserves of starch nourish the seedling. Usually in germination, the breakdown of starch to glucose in the endosperm begins shortly after the seed is exposed to water. Gibberellins in the seed embryo are believed to signal starch hydrolysis through inducing the synthesis of the enzyme α-amylase in the aleurone cells. In the model for gibberellin-induced production of α-amylase, it is demonstrated that gibberellins (denoted by GA) produced in the scutellum diffuse to the aleurone cells, where they stimulate the secretion α-amylase. α-Amylase then hydrolyses starch, which is abundant in many seeds, into glucose that can be used in cellular respiration to produce energy for the seed embryo. Studies of this process have indicated gibberellins cause higher levels of transcription of the gene coding for the α-amylase enzyme, to stimulate the synthesis of α-amylase.

Gibberellins are produced in greater mass when the plant is exposed to cold temperatures. They stimulate cell elongation, breaking and budding, seedless fruits, and seed germination. They do the last by breaking the seed's dormancy and acting as a chemical messenger. Its hormone binds to a receptor, and Ca^{2+} activates the protein calmodulin, and the complex binds to DNA, producing an enzyme to stimulate growth in the embryo.

A major effect of gibberellins is the degradation of DELLA proteins, the absence of which then allows phytochrome interacting factors to bind to gene promoters and regulate gene expression. Gibberellins are thought to cause DELLAs to become polyubiquitinated and, thus, destroyed by the 26S proteasome pathway.

1. Shows a plant lacking gibberellins and has an internode length of "o" as well as it is a dwarf plant.
2. Shows your average plant with a moderate amount of gibberellins and an average internode length.
3.Shows a plant with a large amount of gibberellins and so has a much longer internode length because gibberellins promotes cell division in the stem.

Biosynthesis

GAs are usually produced from the methylerythritol phosphate (MEP) pathway in higher plants. In this pathway, bioactive GA is produced from trans-geranylgeranyl diphosphate (GGDP). In the MEP pathway, three classes of enzymes are used to yield GA from GGDP: 1) terpene synthases (TPSs), 2) cytochrome P450 monooxygenases (P450s), and 3) 2-oxoglutarate–dependent dioxygenases (2ODDs). There are 8 steps in the methylerythritol phosphate pathway: - 1) GGDP is converted to ent-copalyl diphosphate (ent-CPD) by ent-copalyl diphosphate synthase - 2) etn-CDP is converted to ent-kaurene by ent-kaurene synthase - 3) ent-kaurene is converted to ent-kaurenol by ent-kaurene oxidase (KO) - 4) ent-kaurenol is converted to ent-kaurenal by KO - 5) ent-kaurenal is converted to ent-kaurenoic acid by KO - 6) ent-kaurenoic acid is converted to ent-7a-hydroxykaurenoic acid by ent-kaurene acid oxidase (KAO) - 7) ent-7a-hydroxykaurenoic acid is converted to GA12-aldehyde by KAO - 8) GA12-aldehyde is converted to GA12 by KAO. GA12 is processed to the bioactive GA4 by oxidations on C-20 and C-3, which is accomplished by 2 soluble ODDs: GA 20-oxidase and GA 3-oxidase.

Sites of biosynthesis

Most bioactive GAs are located in actively growing organs on plants. Both GA20ox and GA3ox genes (genes coding for GA 20-oxidase and GA 3-oxidase) and the SLENDER1 gene (a GA signal transduction gene) are found in growing organs on rice, which suggests bioactive GA synthesis occurs at their site of action in growing organs in plants. During flower development, the tapetum of anthers is believed to a primary site of GA biosynthesis.

Differences Between Biosynthesis in Fungi and Lower Plants

Arabidopsis, a plant, and Gibberella fujikuroi, a fungus, possess different GA pathways and enzymes. P450s in fungi perform functions analogous to the functions of KAOs in plants. The function of CPS and KS in plants is performed by a single enzyme, CPS/KS, in fungi. In fungi, the GA biosynthesis genes are found on one chromosome, but in plants, they are found randomly on multiple chromosomes. Plants produce low amount of GA3, therefore the GA3 is produced for industrial purposes by microorganisms. Industrially the gibberellic acid can be produced by submerged fermentation, but this process presents low yield with high production costs and hence higher sale value, nevertheless other alternative process to reduce costs of the GA3 production is Solid-State Fermentation (SSF) that allows the use of agro-industrial residues.

Gibberellin Metabolism Genes

One or two genes encode the enzymes responsible for the first steps of GA biosynthesis in Arabidopsis and rice. The null alleles of the genes encoding CPS, KS, and KO result in GA-deficient Arabidopsis dwarves. Multigene families encode the 2ODDs that catalyze the formation of GA12 to bioactive GA4.

AtGA3ox1 and AtGA3ox2, two of the four genes that encode GA3ox in Arabidopsis, affect vegetative development. Environmental stimuli regulate AtGA3ox1 and AtGA3ox2 activity during seed germination. In Arabidopsis, GA20ox overexpression leads to an increase in GA concentration.

Deactivation

Several mechanisms for inactivating GAs have been identified. 2β-hydroxylation deactivates GA, and is catalyzed by GA2-oxidases (GA2oxs). Some GA2oxs use C19-GAs as substrates, and other GA2oxs use C20-GAs.

Cytochrome P450 mono-oxygenase, encoded by elongated uppermost internode (eui), converts GAs into $16\alpha,17$-epoxides. Rice eui mutants amass bioactive GAs at high levels, which suggests cytochrome P450 mono-oxygenase is a main enzyme responsible for deactivation GA in rice.

The Gamt1 and gamt2 genes encode enzymes that methylate the C-6 carboxyl group of GAs. In a gamt1 and gamt2 mutant, concentrations of GA is developing seeds is increased.

Regulation

Regulation By Other Hormones

The auxin indole-3-acetic acid (IAA) regulates concentration of GA1 in elongating internodes in peas. Removal of IAA by removal of the apical bud, the auxin source, reduces the concentration of GA1, and reintroduction of IAA reverses these effects to increase the concentration of GA1. This phenomenon has also been observed in tobacco plants. Auxin increases GA 3-oxidation and decreases GA 2-oxidation in barley. Auxin also regulates GA biosynthesis during fruit development in peas. These discoveries in different plant species suggest the auxin regulation of GA metabolism may be a universal mechanism.

Ethylene decreasing the concentration of bioactive GAs.

Regulation By Environmental Factors

Recent evidence suggests fluctuations in GA concentration influence light-regulated seed germination, photomorphogenesis during de-etiolation, and photoperiod regulation of stem elongation and flowering. Microarray analysis showed about one fourth cold-responsive genes are related to GA-regulated genes, which suggests GA influences response to cold temperatures. Plants reduce growth rate when exposed to stress. A relationship between GA levels and amount of stress experienced has been suggested in barley.

Role in Seed Development

Bioactive GAs and abcisic acid levels have an inverse relationship and regulate seed development and germination. Levels of FUS3, an Arabidopsis transcription factor, are upregulated by ABA and downregulated by GA, which suggests that there is a regulation loop that establishes the balance of GA and ABA.

Gibberellin Homeostasis

Feedback and feedforward regulation maintains the levels of bioactive GAs in plants. Levels of AtGA20ox1 and AtGA3ox1 expression are increased in a GA deficient environment, and decreased after the addition of bioactive GAs, Conversely, expression of AtGA2ox1 and AtGA2ox2, GA deactivation genes, is increased with addition of GA.

Impact on the "Green Revolution"

A chronic food shortage was feared during the rapid climb in world population in the 1960s. This was averted with the development of a high-yielding variety of rice. This variety of semi-dwarf rice is called IR8, and it has a short height because of a mutation in the sd1 gene. Sd1 encodes GA20ox, so a mutant sd1 is expected to exhibit a short height that is consistent with GA deficiency.

Receptors

GA receptors must have four traits: 1) reversibly bind GA, 2) GA saturability 3) high affinity for bioactive GA, and 4) reasonable ligand specificity for bioactive GAs.

Cereal Aleurone Cells

The first reported GA-binding protein (GBP) activity was discovered in wheat seed aleurone homogenates by using the GA-dependent induction of aleuron hydrolytic enzymes. Alpha-amylase induction in aleurone protoplasts is dependent on GA levels, but induction can still occur if GA does not pass through the plasma membrane. The existence of GA perception outside the cell and a plasma membrane GA receptor in aleurone cells can be concluded from these observations. Trimeric G proteins are involved in GA signaling, which supports the previous theory for plasma membrane GA receptors. Four different GBPs have been discovered in the plasma membrane of aleuron cells. Cereal aleurone cells could be unique in GA perception because they are some of the few plant cells that cannot synthesize bioactive GA.

GID1 Receptor

The rice GID1 gene encodes an unknown protein that only has a high affinity for bioactive GAs. In yeast, GID1 binds SLR1, GA signaling represor, depending on GA levels. GID1 is a soluble GA receptor mediating GA signaling. The GA-GID1 complex interacts with SLR1 to transduce the GA signal to SLR1. SLR1 increased the GA-binging activity of GID1 threefold. Arabidopsis has three GID1 homologs that bind GA and interact with DELLAs: AtGID1a, b, and c.

Gibberellin and DELLA Proteins

DELLA Proteins: Repressors of GA-dependent Processes

DELLA proteins act as intracellular repressors of GA responses, DELLAs inhibit seed germination, seed growth, and other GA-dependent pathways, but GA can reverse these effects.

GA-GID1-DELLA Complex

For GA to bind to the GID1 receptor, the C3-hydroxyl on GA must form a hydrogen bond to the tyrosine 31 residue of GID1, which induces a conformational change in GID1 to enclose GA. Once GA is trapped in the GID1 pocket, the lid of the pocket binds to DELLA to form the GA-GID1-DELLA complex.

GA Promotes Proteasome-dependent Degradation of DELLAs

In the absence of GA, many DELLAs are present to repress GA responses, but the formation of the GA-GID1-DELLA complex increases the degradation of DELLAs. The SCF complex is composed of SKP1, CULLIN, and F-BOX proteins. F-box proteins catalyze the formation of polyubiquitin on target proteins to be degraded by the 26S proteasome. The formation of the GA-GID1-DELLA complex is believed to cause a conformational change in DELLA, which enhances recognition of DELLAs by F-box proteins. Next, the SCF complex promotes ubiquitylation of DELLAs, which are then degraded by the 26S proteasome. Degradation of DELLAs allows GA regulated growth to resume. Thus, GA stimulates growth by activating the degradation of DELLAs.

Florigen

Florigen (or flowering hormone) is the hypothesized hormone-like molecule responsible for controlling and/or triggering flowering in plants. Florigen is produced in the leaves, and acts in the shoot apical meristem of buds and growing tips. It is known to be graft-transmissible, and even functions between species. However, despite having been sought since the 1930s, the exact nature of florigen is still a mystery.

Mechanism

Central to the hunt for florigen is an understanding of how plants use seasonal changes in day length to mediate flowering—a mechanism known as photoperiodism. Plants which exhibit photoperiodism may be either 'short day' or 'long day' plants, which in order to flower require short days or long days respectively. Although plants in fact distinguish day length from night length.

The current model suggests the involvement of multiple different factors. Research into florigen is predominately centred on the model organism and long day plant, *Arabidopsis thaliana*. Whilst much of the florigen pathways appear to be well conserved in other studied species, variations do exist. The mechanism may be broken down into three stages: photoperiod-regulated *initiation*, signal *translocation* via the phloem, and induction of *flowering* at the shoot apical meristem.

Initiation

In *Arabidopsis thaliana*, the signal is initiated by the production of messenger RNA (mRNA) coding a transcription factor called CONSTANS (CO). CO mRNA is produced approximately 12 hours after dawn, a cycle regulated by the plant's biological clock. This mRNA is then translated into CO protein. However CO protein is stable only in light, so levels stay low throughout short days and are only able to peak at dusk during long days when there is still a little light. CO protein promotes transcription of another gene called Flowering Locus T (FT). By this mechanism, CO protein may only reach levels capable of promoting FT transcription when exposed to long days. Hence, the transmission of florigen—and thus, the induction of flowering—relies on a comparison between the plant's perception of day/night and its own internal biological clock.

Translocation

The FT protein resulting from the short period of CO transcription factor activity is then transported via the phloem to the shoot apical meristem.

Flowering

At the shoot apical meristem, the FT protein interacts with a transcription factor (FD protein) to activate floral identity genes, thus inducing flowering. Specifically, arrival of FT at the shoot apical meristem and formation of the FT/FD heterodimer is followed by the increased expression of at least one direct target gene, APETALA 1 (AP1), along with other targets, such as SOC1 and several SPL genes, which are targeted by a *microRNA*.

Research History

Florigen was first described by Soviet Armenian plant physiologist Mikhail Chailakhyan, who in 1937 demonstrated that floral induction can be transmitted through a graft from an induced plant to one that has not been induced to flower. Anton Lang showed that several long-day plants and biennials could be made to flower by treatment with gibberellin, when grown under a non-flower-inducing (or non-inducing) photoperiod. This led to the suggestion that florigen may be made up of two classes of flowering hormones: Gibberellins and Anthesins. It was later postulated that during non-inducing photoperiods, long-day plants produce anthesin, but no gibberellin while short-day plants produce gibberellin but no anthesin. However, these findings did not account for the fact that short-day plants grown under non-inducing conditions (thus producing gibberellin) will not cause flowering of grafted long-day plants that are also under noninductive conditions (thus producing anthesin).

As a result of the problems with isolating florigen, and of the inconsistent results acquired, it has been suggested that florigen does not exist as an individual substance; rather, florigen's effect could be the result of a particular ratio of other hormones. However, more recent findings indicate that florigen does exist and is produced, or at least activated, in the leaves of the plant and that this signal is then transported via the phloem to the growing tip at the shoot apical meristem where the signal acts by inducing flowering. In *Arabidopsis thaliana*, some researchers have identified this signal as mRNA coded by the *FLOWERING LOCUS T* (*FT*) gene, others as the resulting *FT* protein. First report of FT mRNA being the signal transducer that moves from leaf to shoot apex came from the publication in Science Magazine. However, in 2007 other group of scientists made a breakthrough saying that it is not the mRNA, but the FT Protein that is transmitted from leaves to shoot possibly acting as "Florigen". The initial article that described FT mRNA as flowering stimuli was retracted by the authors themselves.

References

- Srivastava, L. M. (2002). Plant growth and development: hormones and environment. Academic Press. p. 140. ISBN 0-12-660570-X.

- Öpik, Helgi; Rolfe, Stephen A.; Willis, Arthur John; Street, Herbert Edward (2005). The physiology of flowering plants (4th ed.). Cambridge University Press. p. 191. ISBN 978-0-521-66251-2.

- Weier, Thomas Elliot; Rost, Thomas L.; Weier, T. Elliot (1979). Botany: a brief introduction to plant biology. New York: Wiley. pp. 155–170. ISBN 0-471-02114-8.

- Osborne, Daphné J.; McManus, Michael T. (2005). Hormones, signals and target cells in plant development. Cambridge University Press. p. 158. ISBN 978-0-521-33076-3.

- Roszer T (2012) Nitric Oxide Synthesis in the Chloroplast. in: Roszer T. The Biology of Subcellular Nitric Oxide. Springer New York, London, Heidelberg. ISBN 978-94-007-2818-9

- Elschenbroich, C.; Salzer, A. (2006). Organometallics : A Concise Introduction (2nd ed.). Weinheim: Wiley-VCH. ISBN 3-527-28165-7.

- Kniel, Ludwig; Winter, Olaf; Stork, Karl (1980). Ethylene, keystone to the petrochemical industry. New York: M. Dekker. ISBN 0-8247-6914-7.

- Korzun, Mikołaj (1986). 1000 słów o materiałach wybuchowych i wybuchu. Warszawa: Wydawnictwo Ministerstwa Obrony Narodowej. ISBN 83-11-07044-X. OCLC 69535236.

- Cassells, A. C.; Peter B. Gahan (2006). Dictionary of plant tissue culture. Haworth Press. p. 77. ISBN 978-1-56022-919-3.

- Constabel, Friedrich; Jerry P. Shyluk (1994). "1: Initiation, Nutrition, and Maintenance of Plant Cell and Tissue Cultures". Plant Cell and Tissue Culture. Springer. p. 5. ISBN 0-7923-2493-5.

- Brown, James Campbell (July 2006). A History of Chemistry: From the Earliest Times Till the Present Day. Kessinger. p. 225. ISBN 978-1-4286-3831-0.

- Nomenclature of Organic Chemistry : IUPAC Recommendations and Preferred Names 2013 (Blue Book). Cambridge: The Royal Society of Chemistry. 2014. p. 64. doi:10.1039/9781849733069-FP001. ISBN 978-0-85404-182-4.

- Haynes, William M., ed. (2011). CRC Handbook of Chemistry and Physics (92nd ed.). Boca Raton, FL: CRC Press. p. 3.306. ISBN 1439855110.

- Jeffreys, Diarmuid (2005). Aspirin: the remarkable story of a wonder drug. New York: Bloomsbury. pp. 38–40. ISBN 978-1-58234-600-7.

- Research and Markets. "The Ethylene Technology Report 2016 - Research and Markets". www.researchandmarkets.com. Retrieved 19 June 2016.

- Luckhardt, Arno; Carter, J. B. (1 Dec 1923). "Ethylene as a gas anesthetic". Current Researches in Anesthesia & Analgesia. 2 (6): 221–229. doi:10.1213/00000539-192312000-00004. Retrieved 11 January 2015.

Permissions

Index

www.ingramcontent.com/pod-product-compliance
Lightning Source LLC
Chambersburg PA
CBHW080240230326
41458CB00096B/2740